Pipeline Engineering Monograph Series

Pipeline Geomatics

Edited by:

Steve Adam
Kelsey Davis

ASME Press books are available at special quantity discounts to use as premiums or for use in corporate training programs. For more information, contact Craig Backhus, Special Sales, at backhusc@asme.org

ASME Order No. 802984

ISBN: 978-0-7918-0298-4

Table of Contents

Preface to Pipeline Engineering Monograph Series

The editorial board of the ASME Pipeline Engineering Monograph series seeks to cover various facets of pipeline engineering. This monograph series puts emphasis on practical applications and current practices in the pipeline industry. Each book is intended to enhance the learning process for pipeline engineering students and to provide authoritative references of current pipeline engineering practices for practicing engineers.

Pipeline engineering information is neither readily available from a single source nor covered comprehensively in a single volume. Additionally, many pipeline engineers have acquired their knowledge through on-the-job training together with short courses or seminars. On-the-job training may not be comprehensive and courses or seminars tend to be oriented toward specific functional areas and tasks.

The editorial board has tried to compile a comprehensive collection of relevant pipeline engineering information in this series. The books in this monograph series were written to fill the gap between the basic engineering principles learned from the academic world and the solutions that may be applied to practical pipeline engineering problems. The purpose of these books is to show how pipeline engineering concepts and techniques can be applied to solve the problems with which engineers are confronted and to provide them with the knowledge they need in order to make informed decisions.

The editorial board has sought to present the material so that practicing engineers and graduate level pipeline engineering students may easily understand it. Although the monograph contains introductory material from a pipeline engineering viewpoint, it is reasonably comprehensive and requires a basic understanding of undergraduate engineering subjects. For example, students or engineers need to have basic knowledge of material corrosion mechanisms in order to understand pipe corrosion.

Each book or chapter starts with engineering fundamentals to establish a clear understanding of the engineering principles and theories. These are followed by a discussion of the latest practices in the pipeline industry, and if necessary, new emerging technologies even if they are not as yet widely practiced. Controversial techniques may be identified, but not construed as a recommendation. Examples are included where appropriate to aid the reader in gaining a working knowledge of the material. For a more in-depth treatment of advanced topics, technical papers are included. The monographs in this series may be published in various forms; some in complete text form, some as a collection of key papers published in journals or conference proceedings, or some as a combination of both.

The editorial board plans to publish the following pipeline engineering topics:

- Pipe Material

- Pipeline Corrosion
- Pipeline Integrity Management
- Pipeline Inline Inspection
- Pipeline Risk Management
- Pipeline System Automation and Control
- Pipeline System Design
- Pipeline Geo-environmental Design and Geohazard Management
- Pipeline Geomatics
- Pipeline Project Management
- Pipeline Codes and Standards

Other topics may be added to the series at the recommendation of the users and at the discretion of the editorial board.

The books in this monograph series will be of considerable help to pipeline engineering students and practicing engineers. The editorial board hopes that pipeline engineers can gain expert knowledge and save an immeasurable amount of time through use of these books.

Acknowledgments

We, on the editorial board, wish to express our sincere gratitude to the authors, editors and reviewers for their great contributions. They managed each volume, wrote technical sections, offered many ideas, and contributed valuable suggestions. Supports from the Pipeline Systems Division (PSD) of ASME enabled us to create this monograph series, providing the crucial remainder to the time and expenses already incurred by the editors and authors themselves. We are indebted to the organizing and technical committees of the International Pipeline Conferences (IPC), which have provided an excellent forum to share pipeline engineering expertise throughout the international pipeline community. We were fortunate to have the skillful assistance of the publication department of ASME not only to publish this series but also to undertake this non-trivial task.

Editorial Board

Preface

This book is based on some encouraging feedback from Chapter 8: Geographic Information Systems in the recent book Pipeline System Automation and Control, part of the ASME Pipeline Engineering Monograph Series. As with so many books origins, they start from small beginnings and then build up enough interest until they become a book. In fact, there are several chapters in this book that will make good editions of their own in the future. Besides some general interest in the subject for motivating the book, we felt it is time to put together a broad treatment of the topic of geomatics in pipeline design and operations. The two fields work so well together, it should no longer be a vague and mystical connection.

Our approach to geomatic applications within the pipeline world follows some main focal points of design and operations. These include design and construction, risk and integrity, and regulatory compliance. Other topics addressed where geomatics has had a strong impact and is relevant to our times include applications to environmental stewardship and social responsibility, managing the volumes of data created and maintained in the pipeline business, as well as a look into the future.

The discipline of geomatics suffers from the usual lack of public awareness new disciplines, ideas, and technologies typically face. Geomatics is based on the traditional field of land surveying, and has grown to include techniques and technologies that carry on the surveying tradition of recording *what* (right of way boundaries, pipe welds, in-line inspection data) and exactly *where* (coordinates) it is. Since pipelines are long and linear, they tend to cover large geographic distances for something with such a small footprint. As a result, they are great candidates for geomatic applications since they consist of so many features located in so many different places.

As with many innovations, geomatics has been easily adopted in some cases (e.g. making nice maps for regulatory submissions) and seems painful in other areas (e.g. spatial data management to manage project data). To help pipeline professionals better understand geomatics' contributions to their industry, the authors aim to provide the following:

1. Outline of how geomatics technologies are used in current pipeline practices (design, construction, regulatory, integrity)
2. Address future directions these technologies are leading us to gain a bigger picture and some perspective of the power of geomatics
3. Use this book as a hub to direct readers to other resources in light of the vast array of websites, wikis, and Podcasts
4. Illustrate the multi-disciplinary and collaborative approach taking hold in all engineering fields that generate and utilize volumes of data.

The chapters of this book illustrate many areas where geomatics has helped improve something in the pipeline field and in most cases discuss trends that are taking us into the future. The first chapter introduces geomatics and discusses its impacts in the pipeline value chain using Michael Porter's methodology as well as some perspective on innovation and the connection between geomatics and innovation in pipeline engineering and operations. Chapter 2 introduces the reader to the central concepts of spatial data and data management. It is important to present this chapter early in the book since in forms more foundation for the reader in addition to what is presented in the first chapter.

Chapter 3 begins with a more directed look at pipeline geomatics by demonstrating applications to engineering design and construction and introduces the reader to such innovations as automated alignment sheets. Chapter 4 deals with pipeline integrity, collecting and managing integrity data as well as systems that facilitate integrity and risk programs. Chapter 5 is a timely addition to this book since it addresses our environment and how geomatics technology and practices are helping pipeline companies assess, monitor, and preserve natural areas. Environmental issues continue to be on our minds and impact the public's perception of pipelines and projects. As a result, we look at issues beyond just government regulations to items like cumulative effects and community consultations. Chapter 6 deals specifically with environmental and regulatory compliance and how geomatics (and specifically GIS) have been making permitting and other processes easier for pipeline companies. Chapter 7 deals with geomatics implementations. Part of trying anything new comes the fear of change. We hope this chapter will offer some guidance and security to anyone thinking of implementing GIS on a project or enterprise. The final chapter is a look ahead. Since geomatics is about innovation, this chapter looks at industry and technology trends between geomatics and the pipeline business and what improvements we can anticipate.

As we finished this book, we realized it has fewer pages than we planned. This could be a function of many things but one that we couldn't overlook is that applying geomatics to the pipeline business is in its infancy. As this technology matures we anticipate the applications will grow and mature as well.

Writing this book has been an exciting collaborative effort. We hope you all find something meaningful within these pages and that it provides some information and context for this emerging relationship between Geomatics and Pipeline Engineering.

Steve Adam and Kelsey Davis, Calgary, Canada, August 2009.

Acknowledgment

The editors would like to extend their sincerest thanks to Mike Yoon for his interest in this topic, and his encouragement, support, and reviews writing this book as part of the Pipeline Engineering Monograph Series. In addition, Mike formatted this book which allowed us to focus on finishing it and we are very grateful for that. We would also like to thank the Pipeline Engineering Monograph Editorial Board of ASME for their support and guidance through the writing and compiling. This book is the product of collaboration between many colleagues and it could not have been written so well and so efficiently without their skill, dedication, and character to see this come to life. Our contributing authors are Robert Brook of ESRI, Hernando Buitrago of Inter Pipeline Fund, Murray Rydman of Dynamic Risk Assessment Systems, Samantha Chin of WorleyParsons, Dianne Haley of the Energy Resources Conservation Board, and Glenn Cameron and Marta Wojnarowska of the National Energy Board of Canada. The Rydman family welcomed a new son while Murray wrote his chapter.

Discussing elements of this book with friends and acquaintances is perhaps one of the most enjoyable aspects of writing since it encourages creative freedom and just gets people communicating. We thank the following people for their suggestions, reviews, and valuable information for this project; Perry Peterson and Geoff Clarke of The Pyxis Innovation, Chris Tucker of Point Geomatics, David Thomas of Longhorn Geomatics; George Peric of WorleyParsons, Joanne Sykes of TransCanada Pipelines, and years of conversations with our colleagues at Colt, Nexen, and WorleyParsons. Chapter 5 was developed out of an enlightening conversation with Karen Etherington of TransCanada Pipelines.

Finally, we want to dedicate this book to our wives, children, and families for supporting our efforts and putting up with the distractions writing so often creates. They make doing another book seem possible.

About the Authors

Steve Adam has spent his career as an entrepreneur, scientist, writer, and manager with 15 years of experience coordinating, collaborating, and applying geomatic and information technologies to hydrocarbon and environmental engineering. As president of Adamlabs, he helps clients innovate and improve business practices in a wide variety of industries, notably pipeline, oilfield, and environmental services. Steve has a PhD in environmental engineering and is a big advocate for innovation because we can never improve our condition without change.

Kelsey Davis is a senior Geomatics professional with over 12 years of experience planning, implementing and using Geomatics solutions in a variety of energy and natural resource related sectors. He has fulfilled Geomatics roles at a variety of levels within the pipeline, geophysics, engineering, environmental, upstream oil and gas and geospatial data acquisition sectors. Kelsey holds a B.Sc. in Geography from the University of Calgary and is an Executive MBA graduate from the Haskayne School of Business. He is also Vice Chair of the Alberta Geomatics Group. Kelsey lives with his wife and daughter in the Rocky Mountain Foothills west of Calgary, Alberta Canada.

Robert Brook is ESRI's Global Pipeline and Gas Utility Industry Manager. Over the past 20 years Rob has spent time in many facets of the industry including consultants firms, software and service vendors, and government and nonprofit agencies. He is currently the Chair of ESRI's Pipeline Industry Group, a member of the ArcGIS Pipeline Data Model steering committee, the Vice President of PODS, and on the steering committee of several pipeline conferences. Rob is a regular speaker and contributor to pipeline events in all regions of the world.

Hernando Buitrago has been a member of ASME since 1997. He holds a degree in Mechanical Engineering from the Universidad Nacional de Colombia, and an Executive MBA from the Haskayne School of Business. Hernando has been working in the Canadian Pipeline Industry since 2003, and currently works as Project Manager on pipeline related projects.

Murray Rydman is the Manager of the GIS & Data Management department at Dynamic Risk Assessment Systems, Inc. As a Manager he supervises the GIS group and is responsible for data conversion and integration. Murray has more than 10 years of project driven experience and an advanced technical background in Geographic Information Systems. This experience includes 7 years specialized in pipeline integrity data management and risk assessments.

Samantha Chin is currently a professional engineer working with WorleyParsons Geomatics Ltd in Calgary Alberta, Canada. Ms. Chin has worked in the environmental consulting business for the past 8 years. She is a graduate of the University of Calgary with degrees in Ecology, Geomatics Engineering, and Art History.

Dianne Haley is the GIS Program Coordinator for the Alberta Energy Resources Conservation Board. She has worked in the field of automated mapping and GIS for the past 34 years. She has a BSc in Geography from Queen's University, a Certificate in Information Systems Management from the University of Calgary, a Certificate in Theoretical and Applied GIS from Simon Fraser University, and an MSc in GIS from the University of Huddersfield.

Glenn Cameron is a Senior Pipeline Engineer with the Operations Business Unit of the National Energy Board in Calgary, Alberta. He has a B.Sc. in Chemistry from the University of Manitoba, a M.Sc. in Metallurgical Engineering from the University of Alberta and is a member of APPEGA. He has over 30 years of experience in the oil and gas industry specializing in developing and implementing pipeline integrity management programs.

Marta Wojnarowska is a Systems Project Manager with the National Energy Board. Marta has 16 years experience in Geomatics. She joined the Board as a GIS Analyst in 2004, with a goal to establish a solid foundation for GIS. Prior to NEB, Marta worked for GeoArctic International Services, the Department of Fisheries and Oceans and the University of New Brunswick. She also monitored 2 international Canada/Russia Geomatics Projects for Canadian International Development Agency.

1 Geomatics and Pipelines

1.1 Introduction

The discipline of Geomatics has been supporting the pipeline industry in the form of land surveying for decades. However, the term Geomatics is a fairly modern description of a discipline (1) that has expanded further beyond traditional field surveys to include the elements below. This holistic definition of Geomatics is the one we apply throughout this book.

- global positioning system (GPS)
- geodesy
- aerial and satellite imagery acquisition and processing
- cartography
- web-based and mobile mapping
- geographic information systems (GIS) and
- spatial data management (SDM)

All these geomatics practices listed have innovated aspects of pipeline system design, maintenance, and operations through their ability to generate and deal with data that support business decisions. Innovation is the driving force behind the adoption of geomatic techniques and technology in pipelines as we look to automate processes, do more desktop studies, and store greater amounts of data as it becomes cheaper and is easier to access and store.

The theme that ties geomatics to any application is the concept of <u>location</u>. Fundamentally, geomatics brings the practise of locating, storing, and reporting pipeline-relevant information, based on its location. This is where the common term *geospatial* or simply *spatial* comes from when geomatic technologies and processes are described. Similar to how land surveying would traditionally compile cadastral and as-built information based on where features are located (e.g. right of way boundaries and pipe characteristics later drafted onto alignment sheets) modern geomatics supports the collection (e.g. land survey, ortho-images, bar codes or RFID's on pipe segments (2)), storing (e.g. pipeline integrity database), and reporting (e.g. automated alignment sheets) of geospatial data and information.

Pipelines lend themselves particularly well to geomatic applications. They are inherently geospatial features because they typically cover long linear geographic distances. As a result, geomatics has introduced better ways to collect and organize large amounts of geospatial feature data occurring over large areas and distances. New data collection technologies and more organized ways to manage and report data have enabled us to do more desktop studies, lower risk in cost estimate, centrally manage pipeline integrity, automate drawing production, respond effectively to regulators and the public, address environmental sustainability, and ultimately centralize data on a pipeline project or an operating system.

This book is designed to provide a thorough description of the main themes of geomatics applied to the pipeline industry, namely pipeline design and construction, data management, risk and integrity, regulatory and compliance, and keeping an eye on our environment. We aim offer guidance in these main topics without too much emphasis on timely technologies. There are numerous sources of additional details which we encourage readers to explore since geomatics is about innovation and innovation in content delivery is changing how we all work. For example, there are user groups, online courses, collaborative forums, wikis, and even Podcasts (3). As a result, we are presenting the larger framework of pipeline geomatics, the growing trends, and innovation to the pipeline industry. We hope this supports further enquiry into these applications and successful implementations of a growing and valuable approach to the pipeline business.

1.2 A Short History

Geomatics and pipeline engineering have had a long history together in the form of land surveying. In fact, surveyors were often some of the first people on the job, finding routes, staking out rights of way, drafting drawings showing crossings and land owners, and eventually recording as-built information and registering plans.

Summers spent measuring distances with steel chains, angles measured with inverted optic transits, and winters spent hunched over a hand drafting table characterizes the history of the surveying and mapping industry. Not knowing they would one day be referred to as Geomatics Professionals, surveyors throughout history have observed practices and procedures which have remained largely unchanged until the revolutionary changes in computing and technology late in the 20th century. It is safe to say that the vast majority of pipelines in the world have been designed and built using surveying technology closer to that used when the North American Railroads were built (mid-nineteenth century) than the modern GPS and GIS methods.

The most significant changes in the evolution of Geomatics practices occurred during the 1980's. Electronic distance measuring became commonplace, and affordable computing equipment, and drafting software, saw that maps illustrating special data were no longer prepared on a hand drafting table. Electronic distance measuring meant that a great deal more spatially related data could be collected in the field, thus placing pressure on manual map drafting procedures and encouraging a transition toward automated processes.

These technology advances naturally affected Geomatics workflow on pipeline projects. With an increasingly larger number of Geomatics tools available to them, pipeline designers in the mid 1990's were able to more efficiently plan the design and execution of a pipeline project. Historically, engineers would choose a route and begin an iterative process of surveying the right of way route which, expectedly, was continually being revised due to unforeseen terrain and land owner issues. Pipeline routes today are now more easily selected, avoiding challenging terrain and objecting stakeholders, by using accurate remote 3D views of potential alignments. Many stakeholder concerns, which are often the reason for route revisions, may be satisfied before any field work is initiated. Another benefit to maximizing the efficiency of route design is the reduction of construction related activity and an associated reduction of personal injury incidents on pipeline projects.

There were specific equipment evolutions that dramatically improved efficiency of office and field processes. The change of standard field survey equipment from the common Wild T2 optical transit with external Electronic Distance Measurement (EDM) to today's Real Time Kinematic Global Positioning System (RTK GPS) enables survey crews to monument right of way boundaries much faster and also keep up with the pace of construction, thus allowing a greater accuracy in mapping as-built features such as pipe joints, traps, and line facilities. Through the early 1990's it was common to find ammonia plan copiers in the offices where pipeline drawings were created by hand. Creating and revising hard copies of alignment sheets and construction drawings was obviously a labor intensive manual task. Modern Computer Aided Drafting (CAD) methods and digital drawings put working copies into the hands of engineers, surface land agents, land owners, and approving authorities at a much faster rate using much less effort and labor.

Now that surveyors were collecting more data and processing it much more quickly a need was born to efficiently archive and retrieve the data collected. Geographic Information Systems naturally resulted from the commercial information and equipment technologies that emerged from the 1980's. Faster computers with greater memory seemed to be the initial factor limiting the commercial potential of GIS systems. Today, programming innovation and creativity affect the pace of industry development and it is at this point that this book begins. As we face greater complexity arising from regulatory requirements, a growing demand for energy, and changing political and socio-environmental interests, the pipeline business should take advantage of any innovation that will help it evolve.

1.3 Geomatics and the Pipeline Value Chain – From Planning to Customer Service

1.3.1 Value Chain Analysis

The concept of a Value Chain can be traced back to business strategist Michael Porter in his 1985 book titled, Competitive Advantage: Techniques for Analyzing Industries and Competitors (4). The underlying premise is that for the majority of firms, all activities can be divided into a generalized sequence of value generating activity groupings. Porter generalized those activity groupings as;

Once separated out into the various categories, the activities of the firm can then be more easily analyzed in a systematic manor. The results from this systematic analysis can be fed into improvement programs aimed at finding greater efficiencies, discovering new processes, correcting flawed ones and identifying new markets or channels etc. The end goal ultimately is of course to increase a firm's competitiveness which leads to increased market share and shareholder value.

In the context of a pipeline, it represents a unique element within a value chain. In fact, a pipeline could be considered the primary linkage between different parts of the value chain. A pipeline is the conduit that facilitates the movement of a particular material, be it raw or finished, from one activity to the next along the entire value chain. For example, an Oil and Gas exploration/production company may require a pipeline to get its raw material from the producing well to the processing facility. After refinement, pipelines are then often used or required to get the refined products to market distribution nodes. More pipelines are used to distribute a final product to the end user. None the less, throughout the entire value chain from raw material to finished product and end user, the pipeline forms an integral part of the process.

In considering this, the significance of pipelines as critical infrastructure that supports the entire value chain of a firm becomes readily apparent.

1.3.2 The Pipeline Value Chain

While pipelines form critical infrastructure in support of larger value chains, a unique value chain can also be identified for the pipeline business. Following Porter's template, the value chain for a pipeline firm can be characterized by the following five broad categories of value generating activities;

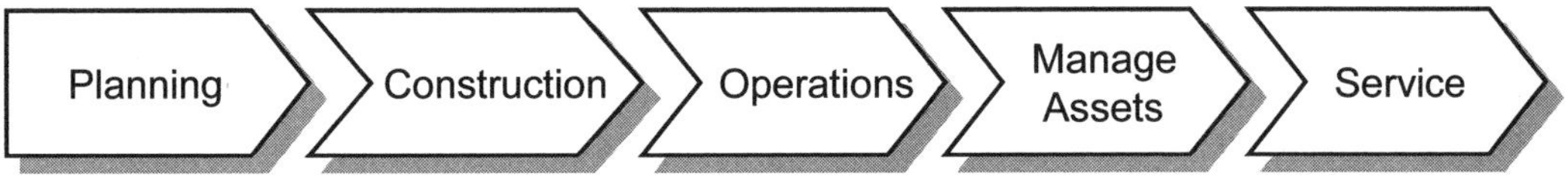

By dividing the pipeline business into these broad categories, it becomes easier to examine the different value creating activities within the pipeline business cycle. When further examined on an individual basis, various opportunities where gains in efficiencies and process improvements can be implemented reveal themselves. By closely examining each of the value creating activities on an individual basis, the opportunities for identifying where Geomatics might be beneficial become easier to recognize.

1.3.3 Geomatics in the Pipeline Value Chain

In a similar fashion to how the pipeline forms the supporting infrastructure for larger value chains, Geomatics provides enabling technology that supports the entire pipeline value chain. From construction through to customer service, the role of Geomatics as a key component linking the entire pipeline value chain is becoming increasingly important.

At the most basic level, pipelines are by their very nature spatial entities. Their reason for being is to transmit a particular medium from one location to another. Geomatics is at its most basic level, a discipline dedicated to the science of examining the spatial world. It is for this reason that pipelines and Geomatics fit so well together.

These concepts coupled with recent advances in computing power, hardware, access to information, ease of software use, greater access to information and improved information acquisition technologies have served to facilitate a natural merger between these two disciplines.

Geomatics technology can be leveraged as a foundation to support more accurate and complete information. Further, the dissemination of information with great ease, to a great number of decision makers at very low cost adds significant worth to pipeline value chain activities.

The essence of where Geomatics provides the maximum value is derived from the impact that it has on information and the systems that manage that information (5). Spatially enabling information systems greatly enhances the value of the information contained within them. Once the spatial component is added to the information, new and unique forms of analysis and reporting can be employed to leverage both the traditional pipeline information in combination with spatial data.

Given the volume of information originally present and with the addition of spatial information, the use of data models in Geomatics workflows allows large amounts of information to be centrally organized and accessible by many. This promotes standards and integrity which in turn leads to better, more consistent decisions (6).

The two main points to remember on the concept of Geomatics within the pipeline value chain are as follows;

1. Geomatics provides a positive impact across the entire value chain.
2. Maximum value and ROI is achieved through the compounding effect of applying Geomatics to the entire value chain (7).

These are key points in that they demonstrate the breadth of value Geomatics offers throughout the entire pipeline business cycle. Probably most profound is the concept that applying Geomatics at the front end of the value chain, and then consistently along it, provides for even greater overall benefit than can be achieved through its application within a single portion of the value chain. To illustrate this, consider a situation where GPS enabled heavy equipment is employed for the construction phase of a pipeline. The cat operator can use GPS enabled sensors and onboard computers to very precisely monitor the movements of his machine. This by itself creates for much greater efficiencies during the operator's work as a single process.

However, consider the impact of preloading all of the alignment sheets, or the ROW details into an onboard computer system in the excavation equipment and then integrating this information with an internet enabled GIS application. Now the activities that are happening in the field on the machine can be monitored in the context of real-time integration with the planning information. Furthermore, reporting on the day to day activities of the machine can be sent back to the planning engineers in the Head office without the need for human intervention. Quicker decisions can be made about the next steps in the process and any potential errors can also very quickly be recognized and corrected. In essence, the pre-construction planning, field construction, and post-construction planning have all been linked together into a continuous information loop. This continuous loop facilitates greatly improved response times up and down the value chain.

To elaborate on the first point, the ability of Geomatics to play a role along the entire value chain is a testament to the diversity and flexibility of the discipline. Further, it is another reason to support the premise that Geomatics fits naturally with the discipline of pipelining. One could even argue that it supports the position that Geomatics *needs* to be applied along the entire value chain if maximum value and ROI is to be achieved.

Another important issue is that the most of the technologies that are being discussed are not new technologies. Although certain elements of them are constantly evolving and being improved upon, the base technologies have been around for several years now. This is important to acknowledge because a common reason put forth in support of *not* adopting new technologies like those Geomatics provides, is a belief that new technologies are unproven, likely to be unreliable and often not worth the cost of implementation (8). This is just simply no longer a valid argument as many of the technologies related to Geomatics have been in use and proven their worth for several years now.

In the following sections a brief discussion of how Geomatics can be applied to the various sections of the value chain is presented. In these sections, select examples for each portion of the value chain will be briefly discussed. Many more examples will be provided throughout the remainder of the book.

1.3.4 Geomatics and Pipeline Planning

Throughout the various stages of planning a pipeline, the use of Geomatics has become an integral part of the process. Initial efforts like understanding where the pipeline needs to start and end and examining potential routes between the two points can be greatly enhanced with the application of Geomatics based technologies like GIS and remote sensing.

Traditional methods of designing, selecting, and maintaining a pipeline route have their roots in manual drafting; a very time and labor intensive process. Increasingly, GIS is being recognized as a valuable tool that can increase efficiencies, while also maintaining the high quality cartographic outputs typically associated with manual drafting. Not only can GIS be used to create and maintain spatial data, design and evaluate potential route locations, and perform analysis or modeling to determine potential impacts; GIS can also be used to produce high quality cartographic outputs in a variety of formats.

Field data collection, often a very labor and time intensive activity, can become very streamlined when Geomatics technologies are utilized. Using tools like GPS enabled tablet PCs with embedded GIS applications and databases while collecting data in the field opens the door for a variety of efficiencies. One example of this would be the directly input of data in the field that is in the proper format for integration into larger enterprise systems at the corporate head offices. Coupled with communication technologies like the internet, near real time, and sometimes real time updating of corporate databases with information collected in the field are easily achievable.

During design, the process of identifying constraints along a proposed route or understanding cost differentials between different options in some instances can be done entirely from the comforts of an office. Advances in computer graphics and processing power have facilitated virtual worlds and visualization without the need for a site visit.

These are but a select a few examples of activities within the planning stages of a pipeline where Geomatics technology has proven very useful in reducing costs and promoting efficiencies. A good description of how Geomatics is being integrated into the pipeline engineering, procurement, and construction management (EPCM) project is found in (9). The author explains that Geomatics is a discipline of similar standing as engineering, procurement, land, environment, and construction management. As a result, it has its own scope which supports other disciplines towards the same goal, be it a cost estimate or detailed design.

1.3.5 Geomatics and Pipeline Construction

The initial pipeline right of way (ROW) pre-planning activities have always involved traditional Geomatics activities like surveying. Identifying and staking out the planned ROW and associated work areas to be utilized during construction as well as marking the work side versus the excavation side of the ROW are all activities carried on by surveyors using traditional equipment like survey chains and theodolites.

Since the introduction of GPS, it has become increasingly common for surveyors to employ electronic GPS equipment in combination with or sometimes as a replacement to traditional tools. Additionally, the heavy equipment used to move earth and excavate might also employ GPS sensors and technology to more accurately perform earth moving activities like cut, fill and grading.

With onboard computer systems, an entire pipeline route can be loaded into interactive software applications that the operator can use to precisely guide his machine's every movement. Reports and logs for daily activity and progress can be quickly generated in an automated routine, then downloaded and sent for review/approval at corporate headquarters. Even real time monitoring becomes a possibility with the proper integration of systems. From the perspective of project management during construction, the benefits become instantly obvious. Furthermore, last minute changes to design and construction parameters can much more easily be accommodated throughout the entire process.

The important thing to recognize is that as processes become increasingly digital, and integrated with supplementary systems like GPS and onboard computers, the speed and accuracy with which information can be collected, integrated, analyzed and then used for decision making increases significantly.

These are but few of the benefits that Geomatics offers the construction portion of the value chain. Chapter 3 (engineering and construction) later in this book will provide a much more in depth look at how Geomatics is increasing value within the construction phase of the pipeline business cycle.

1.3.6 Geomatics and Pipeline Operations

From an operations standpoint, Geomatics has much to offer the pipeline industry. Understanding what is happening and where is paramount in supporting effective operations. In order to get a given product from one end of a pipeline to the other, a number of systems are relied upon to move, monitor and manage the product and its interactions within the pipeline as it flows from one location to another.

In many cases, the demand for energy around the world means that pipelines are required to operate continuously. If one paused briefly to consider the implications of this from an operational standpoint, it is almost hard to comprehend. From monitoring to routine maintenance and upkeep the task becomes an almost insurmountable concept. Even more so when you factor in unforeseen events like equipment failures or line hits. Combine all of these factors with the often large geographic distances over which pipelines are spread out and it quickly becomes quite unreasonable to expect that every location along a pipeline can be monitored at all times.

Often pipeline incidents are not even recognized at their point of origin. A particular event like a line rupture in a remote area might only become apparent after a pressure drop several miles down the line. Even then, the location of the actual problem is not immediately obvious and must be determined by process of elimination.

Databases, computers and scanning systems are all used by pipeline employees to manage pipeline operations. These have been refined over several years to be excellent at providing for an understanding of discrete events and answering the "what" questions. Unfortunately they sometimes do not have the capabilities to always provide insights or answers to the "where" questions. In an effort to mitigate this and properly address "where" questions, Geomatics technologies like GIS systems are becoming increasingly popular as a preferred tool to integrate with the many systems currently utilized to maintain pipeline operations.

There are two basic approaches typically employed when integrating a GIS component into existing pipeline operations systems. One approach is to take an out of the box GIS and customize it to replace the existing systems. This provides a pipeline operations system built on a spatial foundation. The other approach is to add the spatial component to the existing systems. The distinct difference under this scenario is that this creates a geospatially enabled system built on a pipeline operations foundation.

The option of which path to take will be affected by a number of factors, some of which include the age of the existing systems and their flexibility for customization, the culture of the organization, the information technology (IT) and development capabilities and the costs associated with each alternative. Regardless of the path chosen however, both scenarios when properly deployed have proven to provide viable solutions within the context of the operations component of the pipeline value chain.

1.3.7 Geomatics and Pipeline Asset Management

From a dollars perspective, the capital costs associated with constructing and operating pipelines can be very large. Effective management of the resulting assets has a direct impact on the speed and efficiency with which the initial capital outlay can be re-paid. Furthermore the return on investment (ROI) from those assets as a going concern relies, to a large extent, on how well those assets are managed.

At the level of capital required, especially for the larger and longer pipelines, often the cash flow generated from regular business is not sufficient or timely enough. Public companies, where capital for the business as a going concern is raised through selling shares to the public and institutional investors, face great demands to manage assets efficiently. Public and institutional shareholders have a very short tolerance for a lack of return and even less patience for absorbing losses.

Geomatics technology has proven effective for managing assets and mitigating some of the risks associated with large capital investments. As was described earlier, the spatial nature of pipelines makes them exceptionally compatible with Geomatics technologies. A modern GIS, as an example of one of these technologies, provides a framework that hosts all of the necessary information for effective management of assets. Typically, the GIS also provide the interface and tools required for an end user to access all of the information about the assets, and to conduct various analyses as well.

This kind of centralized information management, reporting and analysis structure has proven very effective as a means of managing assets. The key concept with managing any asset is that information is critical. Most comprehensive efforts for information gathering employ the common technique of understanding the answers to the 5 Ws and one H (Who? What? When? Where? Why? How?). This concept is readily applicable in the context of managing pipeline assets.

The value that GIS brings is that it becomes a central repository for the information required to answer the 5 Ws and often provides the tools to figure out the How. One of the most successful applications of GIS for this purpose is its use in pipeline integrity, to which Chapter 4 is entirely dedicated. In addition, Land Management systems, and environmental and regulatory programs (Chapters 5 and 6) would also fall under the asset management category.

Geomatics technology also supports better processes (i.e. how). Very quickly, the value of a centralized repository, reporting and analysis system (GIS) becomes evident (10). How much value can be placed on a system that is easy to use, is universally accessible within an organization and has tools that provide a variety of groups in the organization with the information that they require to focus on what they are there to do. How much value can be placed on making people more efficient? How about the value proposition and competitive advantage that is derived from enabling quicker decision making?

As the above example demonstrates, in the context of asset management it is obvious that properly deployed Geomatics technology can provide widespread benefits across a variety of departments within a pipeline organization.

1.3.8 Geomatics and Pipeline Customer Service

At the end of the value chain is the end customer that ultimately consumes the products transmitted through the pipeline. As has been demonstrated, Geomatics can play an important role in providing value to the previous components of the value chain. At the customer level, there are numerous opportunities for Geomatics to provide value.

One area is in gaining an understanding where customers are. In deciding where and how to build downstream distribution networks both the location and distribution of the customers needs to be considered. Once lines are built to distribute products to the customer, tracking usage and billing becomes important. A modern GIS system is a natural tool for managing large customer databases as many details about a particular location can be related to one another based on the location as a common and unique element.

For tracking usage, meters are typically installed. The exact location on a building, size and technical details for a given meter can be stored and then linked back to billing information. Delinquent accounts can also be flagged and reported on from within such a system.

Immense time savings when field crews need to visit a particular location can also be attained by leveraging such systems. From a safety perspective, pre-site visit details like the presence of an aggressive pet or information about locked access routes etc. can make customer visits more efficient. Having better data has even proven to make employees and customers happier, and decrease operational costs (11).

Extending these corporate GIS systems out into the field with GPS enabled handheld devices or tablet PCs is an excellent way to leverage even greater value out of them. This further helps to facilitate continuity between what is going on at the corporate level and in the field.

Essentially, as is consistent with the other parts of the pipeline value chain, within the customer service segment Geomatics provides a foundation. It is a foundation for more efficient and streamlined workflows, better access to and centralization of information. From these, it has been proven that better, faster decisions can be made which ultimately leads to competitive advantage and increased value (12).

At its very core, geomatics is a technology-based discipline. It's evolution from traditional land surveying has occurred through sophisticated advances in satellites, databases, computer processing power, and lasers. As such, geomatics has elevated many engineering practices through the introduction and application of these technologies through innovation.

1.4 Change through Innovation

As briefly mentioned earlier, geomatics has been traditionally applied to pipelines through land surveying for routing, staking, collecting as-built information, and registering plans. More modern geomatics practices, which still include land surveying, are driven by advances in technology and it is through many of these advances that the pipeline industry has seen innovation.

Any industry that continues to effectively support the demands put upon it by the market and society in general does so by innovating. Innovation is about change, and for a change to be considered innovation it must improve something, like a process, output, thinking, value, or an organization. Innovation can be incremental or it can be disruptive. Incremental innovation is just as it sounds; it happens slowly and is gradual enough that users and the general market have time to adapt to the change, such as the use of spreadsheets. At first they were seen as a tool for accountants, but eventually have become a staple for anyone who uses a computer. Disruptive innovation is exciting, unnerving, and happens with little warning (13). The rapid adoption of web-based social networking and collaborating sites and software (e.g. YouTube, Facebook, Flicker, SharePoint) have caught on rapidly and their full impacts have not yet been felt, nor could we really predict where they will lead us.

Geomatics is no different than other changes that are producing innovation. Certainly, geomatics relies on information technology to manifest these changes, but it is the *spatial way of thinking* that is a big part of the change. This is best illustrated with the concept of *proximity*. Many of the issues that comprise pipeline engineering, safety, operations, and permitting deal with proximity; how close is the pipeline to refiners, HCA's, production, compressor stations, water bodies, cross slopes, rare plant species, etc... Dealing with the issue of proximity is a purely spatial problem and this problem is fundamentally dealt with through geomatic techniques and technologies.

Another area of innovation is centralized data management. Since pipelines are inherently spatial phenomena (i.e. they cover large geographic distances and therefore potentially cover geographically diverse regions) the quantity, variety, and changing nature of features (e.g. the route, encroachment, changing landscapes, and protected areas) require a robust data management system. A spatially enabled data management system simply adds an effective way of indexing the *what* in a database with its *where* component.

When data is well managed and has integrity automating queries and reporting is a natural by-product. Perhaps the greatest examples of pipeline innovation to come out of geomatics are alignment sheet generators. When a sheet is created, everything within a sheet window is pulled from the current database and appears at its geographic location based on a user defined template (Chapter 3 has a thorough description of automated alignment sheets). These are no longer drawings, they are reports based on the current database and can be updated as easily as printing out another report.

The following chapters are designed to highlight the current changes the pipeline industry is feeling through the application of geomatic methods and technologies. We are also uncovering areas where there appears to be some momentum that will carry into the future such as automation, lasers, and Web 2.0. By illustrating the whole picture, current practices, and trends that are appearing on the horizon, we hope to provide pipeline professionals a clearer picture of how they can progress their work with geomatics.

References

(1) Refer to http://en.wikipedia.org/wiki/Geomatics

(2) Hlady, J. and Mondal, S., "Integration of Radio Frequency Identification and GIS for Asset Management", Proceedings of IPC2008, ASME, 2008.

(3) Robert Brook, author of Chapter 2 in this book, is interviewed in an excellent podcast exclusively on the topic of GIS in pipeline applications http://www.esri.com/industries/pipeline/index.html

(4) Porter, M.E., "Competitive Advantage: Techniques for Analyzing Industries and Competitors", Free Press, New York, 1985.

(5) Pearlson, Keri E., Saunders, Carol S., "Managing & Using Information Systems, A Strategic Approach", P. 32, John Wiley & Sons Inc., Hoboken, NJ, 2006.

(6) Pearlson, Keri E., Saunders, Carol S., "Managing & Using Information Systems, A Strategic Approach", P. 4, John Wiley & Sons Inc., Hoboken, NJ, 2006.

(7) Adam, S. and Hlady, J., "Data is an asset that should be managed", Proceedings of IPC2006, ASME, 2006.

(8) Carr, Nicholas G., "IT Doesn't Matter", Harvard Business Review, May 2003.

(9) Parker, D., "A new role on the pipeline EPCM project", Proceedings of IPC2006, ASME, 2006.

(10) Yoon, M., Warren, C. B., and Adam, S., "Pipeline System Automation and Control", Chapter 8 – Geographic Information Systems, Pipeline Engineering Monograph Series, pp 325-375, ASME, N.Y., 2007.

(11) Redman, Thomas C., "The Impact of Poor Data Quality on The Typical Enterprise", Communications of the ACM , Vol. 41 No 2, February 1998.

(12) Tomlinson, Roger, "Thinking about GIS Geographic Information System Planning For Managers", P. 6, ESRI Press, Redlands, California, 2003.

(13) Von Hippel, E., "Democratizing Innovation", P 216, The MIT Press, Cambridge, MA, 2005.

2 Spatial Data and Spatial Data Management

2.1 Introduction

Spatial data, data that includes geography, is one of the most undervalued assets owned by a pipeline operator. It is easy for an organization to assess the value of a valve or a pump station, or to measure the contribution of an employee. While these physical asset calculations are straight forward, assessing the value of spatial data is more complicated (1) (2).

Data is expensive to generate, analyze, report, and store. While this may be perceived as an initial obstacle, it is really a very powerful investment. Once data is available and integrated into workflows, companies can see its real worth. Good data reduces the risks associated with every decision and helps a company make better, more educated choices.

Every company has information from a multitude of sources including as-built survey, inline inspections, satellite imagery, and commercial vendors. This information is stored inside boxes in regional field offices, on the desks and computer of employees, and in filing cabinets or corporate storehouses and servers. While data is prolific, uncontrolled data access can be dangerous as only current data has real value. In fact, working with out-of-date information can be just as dangerous as not having information at all—it creates an illusion of being well-informed.

The conundrum for many companies is how to ensure data is accurate, and how to make certain that people are using the correct data. Those two questions are the driving force behind what is referred to as a geospatially enabled enterprise data storehouse where all data is geographic, the information is widely available and there are establish maintenance processes to ensure accuracy.

Once spatial data is available, it plays a vital role in all facets of an operator's business including maintenance, operations, integrity and risk analyses, regulatory compliance, and legal processes. Valuable information, not just data, must be made available to all employees.

Before this can be done it is important to know what defines spatial data. What are key details and specifications about the data? Which management methods will be used? How will the data be integrated? The remainder of this chapter is an introduction to these topics.

2.2 Spatial Data Types

The first requirement for effective data management is a good understanding of the data itself. For each of a pipeline operator's activities, there is specific necessary information. The operator must be able to understand all the relevant data.

Typically an operator confronts a “bucket” of data that contains all available information. This information can be organized in many ways. Here, we will group it as corporate data, operational data, and base data. This data stack is illustrated in Figure 1.

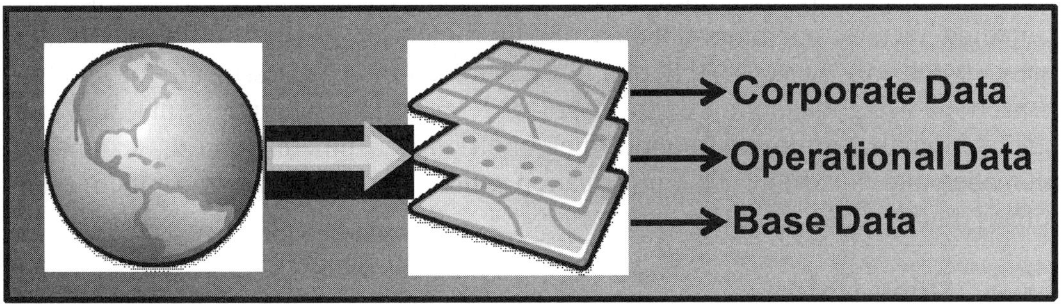

Figure 1 – data stack showing corporate, operational, and base data

2.2.1 Corporate Data

The term corporate data is often used to describe field assets owned by an operator, including leased or owned property; inline appurtenance; design specifications; and the actual pipeline. Land data is maintained by the land department and may be stored as spatial or non-spatial records.

Pipeline and appurtenance information often comes from field or construction survey files. These files are the product of as-built survey data collection activities, completed by internal field staff and external contractors. In the case of acquired or purchased assets, the files may come from an unknown source. This survey data defines the three dimensional location of the pipeline and related infrastructure, and acts as a permanent record of what was assembled including all the necessary deviations from the original engineering design. Essentially, the as-built file contains exactly what actually went into the ground and where it is located.

Corporate data can also refer to information created during the engineering design and construction process. During the preliminary stages of pipeline design, an organization completes appraisal activities including right-of-way and environmental surveys. In more complicated designs, this may also include soil- and surface-geology surveys. During the lifespan of a pipeline, operators will also complete survey related to property, house occupancy counts, repairs, and maintenance activities.

2.2.2 Operational Data

Operational data is information related to, or created during, the operation of a pipeline. This includes transmission records or tickets that document pipeline flow through; operating parameters such as pressure or temperature; cycle information; and any other measurement captured by the SCADA system.

A second type of operational information is integrity records. As a pipeline ages, natural forces decay and corrode the materials. To determine the impact of degrading factors, operators conduct pipeline integrity surveys. While there are many types of surveys—external inspection, direct assessment, or inline inspections using magnetic flux leakage (MFL) or ultrasonic equipment—they all generate detailed information about the condition of a pipeline and the extent of anomalies and corrosion. This information guides the operation of a pipeline or defines maintenance activities.

2.2.3 Base Data

While survey data documents the pipeline and its condition, it is complimented by base data that describes the regional geography and characteristics of a pipeline's operating environment. A pipeline organization may locate freely available information, such as GIS data from the USGS clearing house (3), or purchase necessary data from external vendors. This may include regional cadastre or property fabric, property or structure occupancy information, detailed transportation layers, geological or geographic data, waterways or hydrographic datasets, etc.

Most pipeline companies also maintain a collection of satellite images, aerial photographs, and helicopter video data. This information is often used as the background for maps and alignment sheets, but it can also be used to generate additional base data such as property boundaries, vegetation, roads, or any other visible geographic characteristics.

2.3 Further Defining Data

While classifying information into thematic types is helpful, it is also important to define data as spatial, attribute or non-spatial. Essentially, spatial data is constructed from two pieces: the absolute or relative geographic component, and the characteristics or information about it. Let's begin by looking at the spatial element.

2.3.1 Spatial Data

While one could argue that any piece of data is inherently spatial—as data can always be related to something with a location—spatial data is located somewhere.

Real spatial data has a location defined using a geographic coordinate system. This is where a three-dimensional spherical surface is used to define locations on the earth, and is either imbedded into the file (as with GIS data) or stored as X and Y co-ordinates.

While all pipeline data could be linked to a location, only information with a location is referred to as spatial. Base data is always spatial in nature, as is most corporate data. Only some operational data is spatial. In order to differentiate between spatial and non-spatial data, and to understand the nuances within geographic information, we must first appreciate how a location is established.

2.3.2 How Do We Define Location?

The foundation of a geographic coordinate system is where a particular point or location is referenced using a measurement system with longitude (the X measure) and latitude (the Y measure). Longitude and latitude are based on angles measured from the earth's center to a point on the earth's surface. The location is shown in decimal degrees, or degrees, minutes and seconds (examples are shown in Figure 2). Data defined using a geographic coordinate system is displayed as if a degree is a linear unit of measure. The shape of the earth is represented using a combination of a spheroid approximating the shape of the earth, and a datum defining the position of the spheroid relative to the center of the earth. As shown in Figure 2, the latitude/longitude grid is draped over this spheroid. This is a very common form of measurement. You will find it on any child's globe.

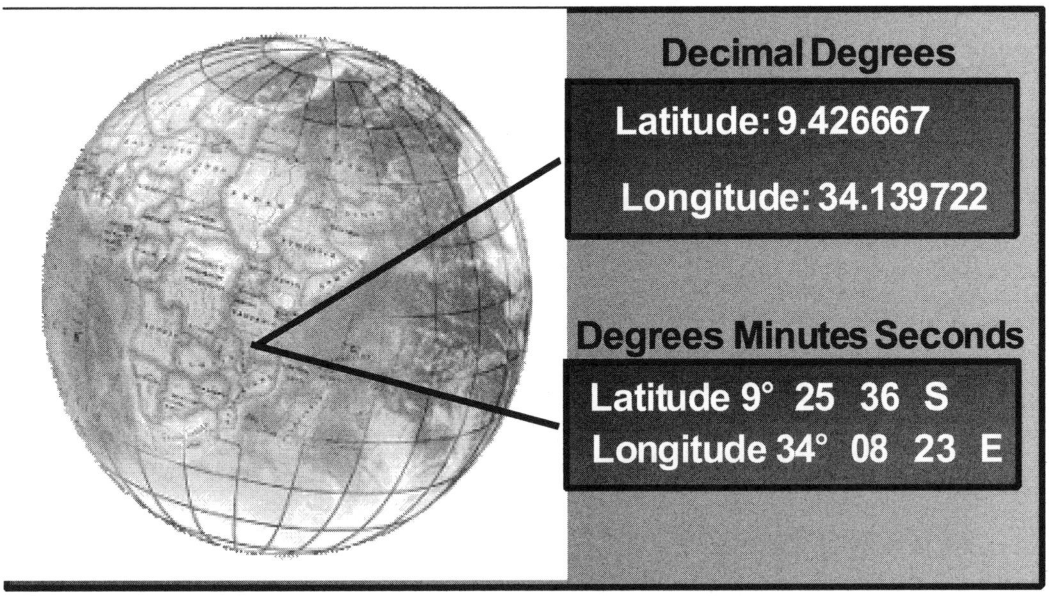

Figure 2 – The Earth with the Latitude and Longitude Grid and Measurements for Lake Malawi

Although longitude and latitude can locate exact positions on the surface of the earth, they are not uniform units of measure. As you proceed toward either the North or South Pole the circumference of the earth shrinks, and the distance represented by a degree of longitude decreases until it finally equals zero at the poles. Because degrees of latitude and longitude don't have a standard length, you can't measure distances or areas accurately or display the data easily on a flat map or computer screen. While this is not a significant problem for small scale mapping, it is significant for most corporate agencies, including pipeline operators who work with larger scale projects.

To solve this problem most companies reference and map their data using a projected coordinate system. Although it is based on a geographic coordinate system, a projected coordinated system is a flat, two-dimensional representation of the earth. It is still based on a spheroid, but it uses linear unit coordinates so that calculations of distance and area are easily done. The latitude and longitude coordinates are converted to X, Y coordinates on the flat projection where the X coordinate represents eastward direction and the Y coordinate represents northward direction. This is shown in Figure 3.

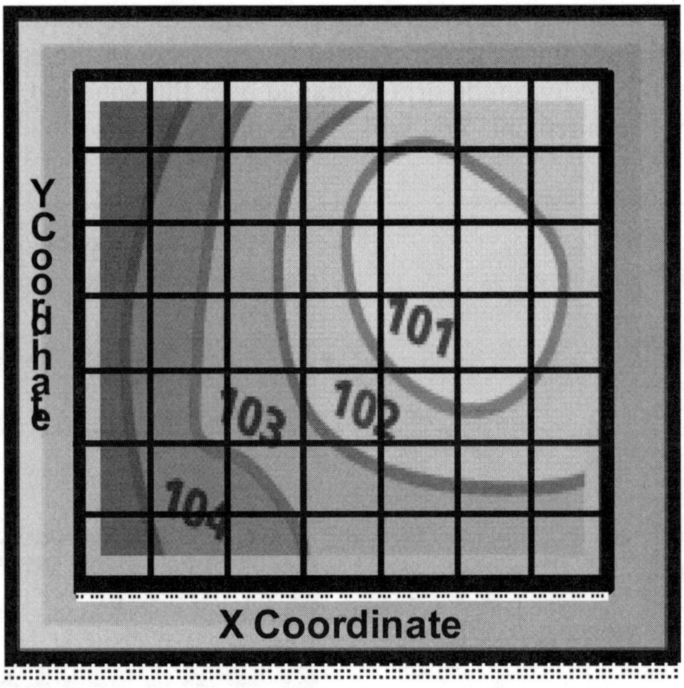

Figure 3 – X and Y are the Foundation of a Coordinate System

Mathematical formulas are used to convert a three-dimensional geographic coordinate system into a two-dimensional flat projected coordinate system. The transformation is referred to as a map projection, and there are many types including conic, cylindrical, and planar. Depending on the projection used, different spatial properties will appear distorted. While this can result in some inaccuracies, it is much more suitable for the large-scale mapping activities performed by pipeline operators. The most prolific example of a projected coordinate system is the Universal Transverse Mercator (UTM) which uses the Transverse Mercator projection.

2.3.3 Raster and Vector

Using a geographic or projected coordinate system, spatial data is visualized in two different ways: as rasters or vectors and shown below in Figure 4.

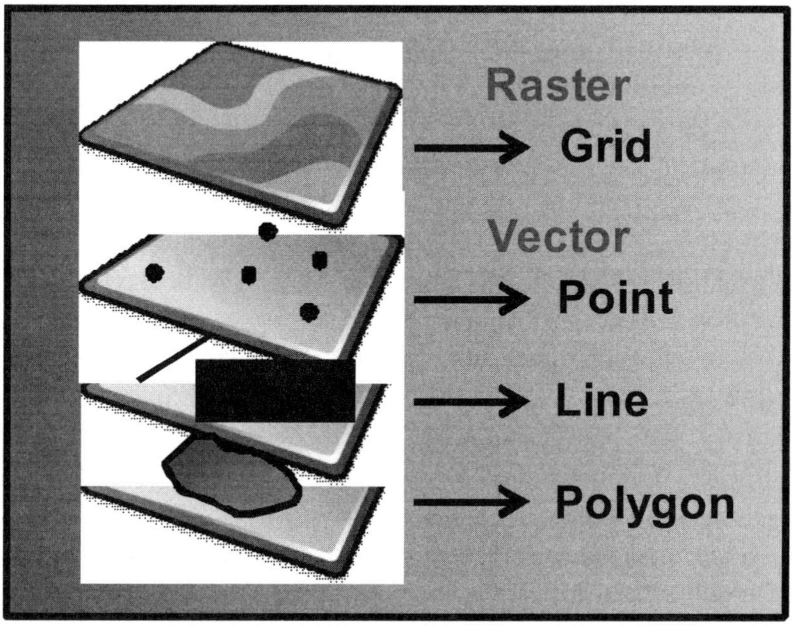

Figure 4 – Raster and Vector Data

Raster data is a grid-cell data structure where the geographic area is divided into cells of equal size identified by rows and columns. Raster data is essentially a gridded representation of some type of feature such as vegetation, soil type, or elevation where each individual cell has a specific measure. For example, in an elevation raster, each cell has a specific vertical measure.

Vector data represents geographic information using points, lines or polygons. The basis of all vector data is a point. A point can be used to represent a piece of data such as cathodic protection anode or a pump, or sequential points (called vertices or nodes) can be connected to form a line or linear segment. A polygon is a created when linear segments connect end to end until the feature is "closed" and an internal an area is created. While each point or vertex consists of an X and

Y coordinate, only points, lines, and polygons have attributes (see below for a description).

2.3.4 Precision and Accuracy of Spatial Data

Geographic location is represented as coordinates. These coordinates are shown as a number. As with other mathematical calculations, for example pi, the more precise the measurement the larger the number of decimal places. Therefore the precision of the geographic data is defined by the number of decimal places it has. The larger the number of decimal places, the higher the precision.

While precision is important, greater precision has drawbacks. It is much more complicated to deal with and manage, requires more hard drive space to store, and system/analytical performance decreases as precision increases. An operator must create a balance between precision and performance to have an effective geographic analysis environment.

Data precision can be established in many ways. First, the surveyor or computer process that created the data may have only calculated to a certain number of decimal places. Second, if the information required human transcription or input, that individual may have established the precision. In addition, as most usable geographic information is digital, data precision may have been established by a system administrator.

While data accuracy can be impacted through data modification and human error, it is established by the equipment or process used to create it. For example, the accuracy of survey data is defined by the technology used to collect the information, but it can be impacted by human error in the set up of the equipment.

While raster accuracy is subject to the same controls as vector data, spatial resolution is an additional factor for consideration. Spatial resolution refers to the dimension of the cell size where the dimension represents the area covered by a single cell.

A higher spatial resolution indicates that there are more pixels per unit area, and the higher the resolution the greater the visible detail. High-resolution data is always preferable because it allows you to identify features that are not visible in lower resolution data. Since high-resolution data is more expensive and much larger, most pipeline operators use it only for the rights of way.

2.3.5 Vertical and Linear Measures

Spatial data can have two other types of measures: a vertical measure (Z) and a linear measure (M) which are both illustrated in Figure 5. In vector data, the Z measure has its own vertical coordinate system which defines the origin for height or depth values. The vertical coordinate system has a vertical datum, a linear unit of measure, an axis direction, and a vertical shift. The vertical unit of measure is always linear, for example feet or meters. Raster elevation surfaces (digital elevation model) are also available where each cell has a specific elevation. This vertical element is particularly important to pipeline operators as it plays an integral role in the calculation of slack chainage, depth of cover, and volumetric calculations.

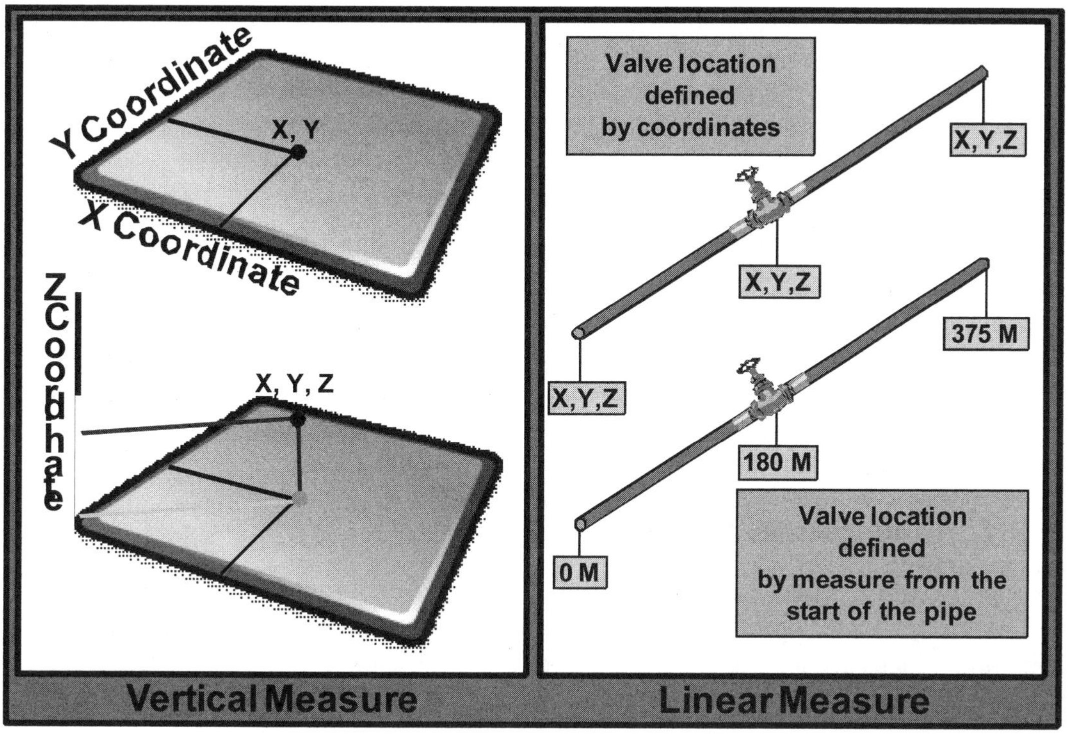

Figure 5 – Vertical and Linear Measures

Linear vector data can also have measures. This is where a measure is interpolated for a specific location along linear features using the beginning of the line as the starting point. We will return to this topic later in the chapter when we discuss linear referencing.

2.3.6 Attribute Data and Non-Spatial Data

Now that we understand the geographic component of spatial data, let's look at the second element: attribute data. While everything occurs in a location, we know something more about it. For example, we know a valve exists at an X, Y coordinate, but we also know its type, model number, who installed it, when it was installed, an inventory ID number, etc. Within a raster data set each cell represents something. All of this additional information is called attribute data. These characteristics can be quantitative and/or qualitative in nature and are often referred to as tabular data. Almost all spatial data has attributes.

Pipeline companies maintain a great deal of spatial data as CAD Drawings, GIS data, and tabular information with X and Y coordinates. Additionally, they have a significant amount of operational data that is non-spatial. Non-Spatial data is very similar to attribute data. It is often in a tabular form but while it can be related to a geographic feature, the spatial information is not present. An example of this might be SCADA measurement or maintenance records. They occur at a location in the system but the geographic component is not stored along with the data. While this information is not geographic it can be "spatialized" by establishing relationships using database joins or relates to spatial data.

2.4 Data Specification

So now we have an understanding of the kinds of data that a pipeline company may have and how it is spatially referenced. Let's investigate how spatial data becomes connected to a pipeline. Pipeline operators manage spatial information about linear systems. It is easy to visualize the pipeline as a line, but how does inventory and incident data that occurs along the linear networks get tied to its specific location on the pipeline? This is accomplished using measurements and linear referencing.

2.4.1 Linear Referencing

Normally, places are located on the earth using a geographic coordinate system, and latitude and longitude or X and Y coordinates. For linear features like a pipeline, places are also located using a relative position along a linear feature.

Essentially there are two methods for identifying the spatial location of these elements: the geographic location and a relative position related to another feature. Some examples are a compression station 8 miles from the start of a pipeline; a valve that occurs 1080 meters from a launcher receiver station; a dent positioned 1247 feet from a metering station. The method of referencing locations by measurements along a linear feature or route is called linear referencing.

Some equipment is located at specific points, like a tee or a weld, which are called point events, defined by a measure along a route. Other features, such as coating or a section at a certain operating pressure, are located along a section of a pipeline route. These are called linear events and are defined by a starting point (from-measure) and ending point (to-measure) along a route.

In the pipeline industry, linear referencing is often referred to as stationing. There are many types of stationing. Absolute stationing defines measurement by absolute location. Slack chain stationing takes into account vertical variations. Engineering stationing is a complex form of measurement using mathematical equations to account for changes in pipeline length caused by construction and rerouting. Regardless of the type of stationing used, any point along a pipeline can be uniquely identified. Linear referencing is essential for collecting and storing information involving facilities, physical or inline inspection data, residency and regulatory compliance information, risk assessment, and geographic information such as environmentally sensitive areas, cultural features, surface geology, political boundaries, hydrography right-of-way boundaries, or pipeline crossings.

2.4.2 Topology

Just as there is a direct relationship between a pipeline and its features, there is a relationship between different types of data based on physical association. Take, for example, a pipeline right of way and the associated properties held by land owners. In the real world we can see that something, such as a fence, separates the properties from the right of way. In the computerized world of spatial data, something needs to define how point, line, and polygon features share geometry. This is called topology.

Topology characterizes and enforces data integrity rules such as, "there should be no gaps between polygons;" and it supports topological relationships such as feature connectivity. In other words, topology ensures that physically associated data forms a consistent and clean topological fabric. In the case of our property example, topology ensures that the polygons representing properties and right of way stay connected, thereby properly embodying the real world.

2.4.3 Pipeline Geomatics Standards

The topic of geomatics standards for the pipeline community is nebulous. The geomatics industry itself has many standards. One group of standards relates to the presentation, interoperability, and encoding of data. Examples of these are the standards created by the Open Geospatial Consortium (4). There are also evolving global standards for geospatial information such as the 45 published standards by the International Organization for Standardization's Technical Committee 211 on Geographic Information (5). In addition, many countries have established national standards for geographic data.

The pipeline industry has its own standards such as those established by American Society of Mechanical Engineers B31.8 for gas pipelines or B31.8S gas pipeline integrity. There are also the corrosion control and prevention standards promoted by the National Association of Corrosion Engineers, and the database standards promoted by the Pipeline Open Data Standard. Each of these has some bearing on what data is created, the attributes that are maintained, and how a pipeline operator stores and transfers spatial data. But, of paramount importance are the standards pipeline operators create for themselves.

To ensure universal data accessibility and consistency, every operator needs to establish personalized data standards. These standards allow variations during workflows or specialized processes, but stipulate requirements for permanent data. The development starts with defining the organization's standard data format; as well as a universal coordinate system, datum, and spheroid. It requires that mapping standards are created indicating the projection, scale, symbology, and color patterns. These standards should extend from presentation and graphical elements down to the pipeline data model—the foundation of many operators' mission-critical business functions such as pipeline construction, maintenance, and integrity management programs or regulatory compliance. Standardization will provide consistency and eliminate the need for costly data transformation.

2.4.4 Data Models

One of the main methods for spatial data standardization is the use of an industry pipeline data model. The model provides both the structure for data storage, maintenance, and use; along with the environment for solutions, interoperability, and enterprise integration. Data models are essential to the pipeline industry as they allow users to store, integrate, and analyze data faster and more effectively. This need becomes more significant as pipeline operators address an expanding number of operational requirements, implement new technologies, and manage the resulting volumes of information.

More and more operators desire a data system that is stable, allowing a pipeline GIS to grow into an organization-wide resource that can be integrated with other enterprise systems. It is obvious that an operator's pipeline data model has become the foundation for a new type of mission-critical system. For a GIS to be effective, the model that is selected must suit the specific needs of the operator.

When implementing any type of data storage system, an organization also needs to understand similarities and differences between data templates, data models, and data standards. While each of these options can provide an operator with different opportunities, the selection of the appropriate type should be based on the organization's requirements, work flows, existing processes, and culture. The proper selection will allow an organization to remove future hurdles and integration hardships. The industry has several models to choose from but the two most commonly implemented are the Pipeline Open Data Standard (PODS) (6) and the ArcGIS Pipeline Data Model (APDM) (7).

The PODS model was developed and is maintained by the PODS Association. The membership of the association includes pipeline operators, solution vendors, consulting companies, and regulatory agencies. The PODS model is a standard and is meant to be implemented in its entirety. It stores pipeline data required to meet key business drivers such as regulatory compliance, integrity management, facilities and operations management, data collection, land, and others needs as defined by the operator community. The data model is continually expanding to include any and all data associated with the pipeline industry. One of the primary principles of the PODS Association is that all data models they create and data-exchange standards they work with remain "open."

Currently the PODS model is relational and built to run using a Relational Database Management System. Other than storing linear referencing and coordinate information in tables, there is no inherent spatial functionality. Vendors and software companies develop different spatial solutions and methods for spatializing the data. In the near future PODS will be releasing two spatial versions of their model, a geodatabase model and one designed for Oracle Spatial.

The ArcGIS Pipeline Data Model is a free ESRI geodatabase created for use with ESRI's ArcGIS products. The APDM was initially designed from existing published data models. This content was expanded by a group of industry representatives to meet the needs of gas and liquid transmission pipelines. The model includes a sampling of features generally found in the majority of pipeline companies, but was tailored to ensure that specific items such as integrity, pipe inspection, high-consequence areas, and risk analysis were more comprehensive. APDM was not designed to be a model; rather it was created as a template that a pipeline operator could use as a starting point. As each organization has different nuances to their business they would extend and refine the model as they see fit by adding features or modifying existing features. As a GIS-based model, APDM is inherently spatial and does not require additional spatialization tools.

2.5 Data Management

Let's assume for a moment that we work for a pipeline operator with organized data. They have spatial and non-spatial data, and have migrated the necessary data into one of the standard data models. So how does a user see the data they have?

While tabular information can be viewed using a database or spreadsheet application, it is not conducive to spatial visualization. It is important for the user to see spatial relationships between data, and apply a geographic approach to business decisions. This is done using a geographic information system (GIS).

A GIS is a software application used for capturing, managing, analyzing, and displaying geographically referenced information. A GIS allows us to observe, comprehend, question, interpret, analyze, and visualize data. These processes reveal relationships, patterns, and trends within and between datasets. A GIS lets us document these relationships and visualize spatial data in the form of maps, globes, reports, and charts. Most often we relate GIS with a map view of spatial data. GIS can also be used to investigate a tabular or database display, and to visualize or create higher level analysis in a model or geoprocessing form.

Once a user is able to visualize spatial data, the next big challenge is to keep this information current. How is this done? It is completed using the GIS and a distributed process referred to as spatial data management.

2.5.1 Spatial Data Management

Many different workflows may be combined to achieve the goals of spatial data management. Each operator separates these tasks between different departments or teams; and it is rare to find two organizations that do it identically. Let's return to our initial pipeline data categories and discuss upkeep requirements for each type of data regardless of the department responsible for its maintenance.

2.5.1.1 Base Data

While this is not always the case, most pipeline operators do not maintain their base data. Instead they have established relationships with external providers who create and maintain these data sets and deliver them on a predefined schedule. This includes companies such as i-cubed, First American, White Star, Penwell, Tele Atlas, NAVTEQ, Microsoft, Google, or ESRI, just to name a few. This data can be provided in many forms—a database, files such as shape files, or a Web service. In the case of a database or files, the operator is tasked with uploading new data. Web services do not have data loading overhead and are quickly becoming the standard for commercial base data acquisition.

Base data can also be acquired from government agencies through the American National Geospatial Program, Canada's GeoBase, the USGS, and numerous other federal, state, or provincial agencies. It can also be accessed through numerous data portals such as geodata.gov or the geography network. Regardless of the origins of the data, base data is not commonly spatial information that a pipeline operator has to maintain.

2.5.1.2 Operational Data

As with base data, operational data is not commonly maintained by GIS users. Some operational data, such as inspection data, is never physically updated or replaced unless a pipeline is repaired or taken out of service. For example, inline inspection records are not updated. The information is used in pipeline analyses until a new inspection is completed and then the new data is added to the analysis. Other data sets are maintained by external applications such as SCADA, SAP, or Maximo, and the GIS user's focus is on accessing/spatializing the data to take advantage of changes made by the parent system.

2.5.1.3 Corporate Data

There is a great deal of corporate data that is spatial and requires a maintenance process within a GIS environment. Often this is referred to as the asset or facility maintenance system. Maintenance solutions are used to create, remove, update, and manage the location and attribute information of spatial features in the pipeline database. Data maintenance can occur in two locations: the office and the field. Each location requires a different type of solution but an example of how they can work together is shown is Figure 6.

Office solutions focus on supporting and facilitating workflows. Examples include inputting new engineering designs; updating and correcting existing data based on field inspections; adjusting data to account for maintenance and repair activities such as rerouting, inserting clocksprings or sleeves, or pipeline replacements. Office solutions exist as extensions to a GIS solution, or as custom software products based on a GIS. In either case they take advantage of the GIS's core editing and linear referencing technology.

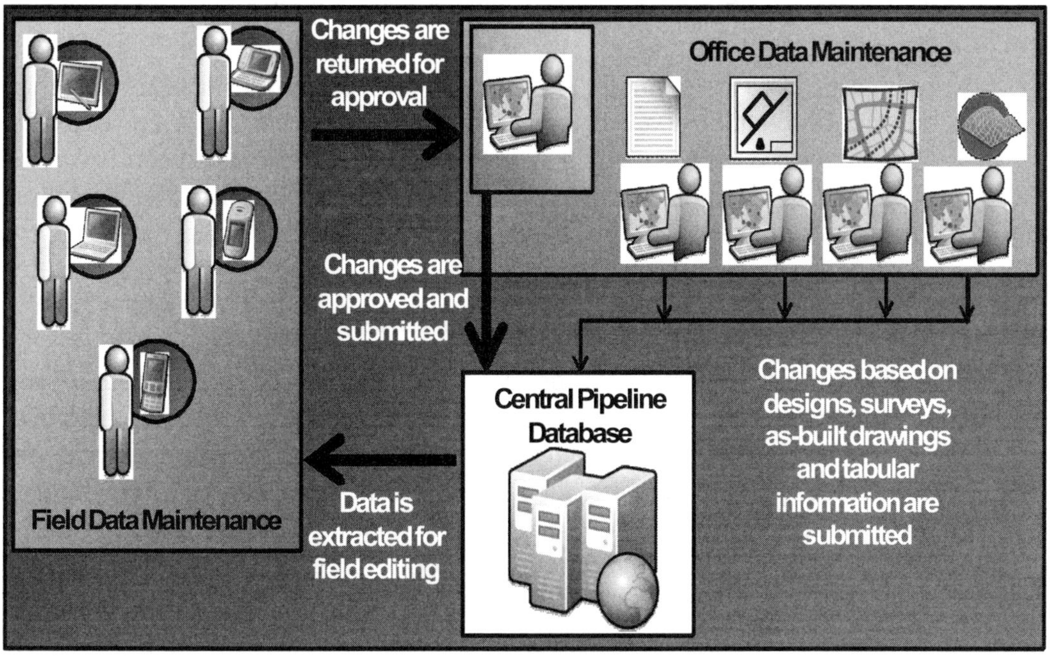

Figure 6 – Example Maintenance Architecture

The second type of asset maintenance solution is for operations and maintenance groups who are responsible for inspecting and maintaining pipeline field infrastructure. The solution facilitates activities required by regulations, integrity management plans, and best practices; or they can be used to conduct reactive work based on emergency or priority situations. Scheduled activities include things such as internal and external tool-based inspections, direct assessment and dig reporting, and any pipeline maintenance (valve servicing) or asset replacement activity. This work covers all pipelines, online and offline appurtenances (valves, cathodic protection, etc.), and facilities (pump stations, processing plants).

When conducting operations and maintenance activities, field staff is responsible for completing inspection forms or reports that document critical information. This includes asset surveys updates, integrity dig reports, foreign line crossing reports, site condition assessments and routing forms. Traditionally, these inspections have been recorded on paper, but GIS-based field solutions are quickly replacing paper. These field data collection solutions are a form-based computer or mobile devise application that is used to collect the information traditionally put on paper. These applications closely model the traditional workflow and can be extended to include appliances such as digital cameras, bar code scanners, or laser range finders.

The field applications work with, or extract from, a version of the pipeline database. Data is updated with field values and then returned to the office. Once returned, the new information is migrated back into pipeline database. Newer field solutions are using wireless technology or broadband cards to transmit changes in real time back to the office.

Pipeline operators always maintain an approval process to ensure that changes made to the pipeline database are correct. Edits are submitted and then reviewed by a peer or supervisor who is tasked with the approval process. If the reviewer ascertains that the modifications are valid, the change is posted to the database, modifications are committed, and the edit is complete.

2.6 Enterprise Data Integration (Image)

Spatial solutions rarely live in isolation. We have already described operational data that lives outside the spatial environment. And, every pipeline organization has implemented numerous key enterprise information technologies. In some cases this includes client relationship management systems (CRM), human resource management (HRM), enterprise resource planning (ERP), and document management systems (DMS). For others it is work order management (WOM), asset management systems (AMS), and SCADA.

Regardless of the type, the systems were selected because of their ability to optimize operations. While the complexity of the enterprise playing field is fully realized by IT staff, users expect instant access to all business data that exists within the enterprise regardless of the system. Accessing this data requires that disparate applications be connected to a larger, integrated solution. This integration is usually achieved through an enterprise integration strategy.

Architecting an enterprise integration strategy is a complex task. There are many conflicting business drivers, and whether the architecture was in fact a right choice is not known until it is put to the test. This topic extends far beyond the concerns of the GIS/spatial data users.

Over the past five years, organizations have been diligently working to develop integrated, multi-departmental enterprise architectures specifically designed to address specific business needs. The strategies have done an excellent job of providing broad access to geospatial data, creating a common infrastructure upon which to build and deploy GIS applications, and exposing organizations to cost reductions through economies of scale.

A common link between all of these enterprise systems is that the information they manage have a relationship to geography; but few of these applications have access to the geography. In parallel, GIS users are maintaining geography at a great cost. Integration projects are making business data available to GIS users, but are also exposing geographic information and GIS functionality to other enterprise solutions. This shift provides a spatial context across enterprise systems and makes geoprocessing capabilities available to any application that may require it.

The two most common methodologies for integrating enterprise systems are direct integration and the services oriented architecture (SOA). Direct integration requires that point-to-point connections be created between databases. This is a proven methodology, but often these connections are project-based and the volume of them makes their management a complicated and difficult activity.

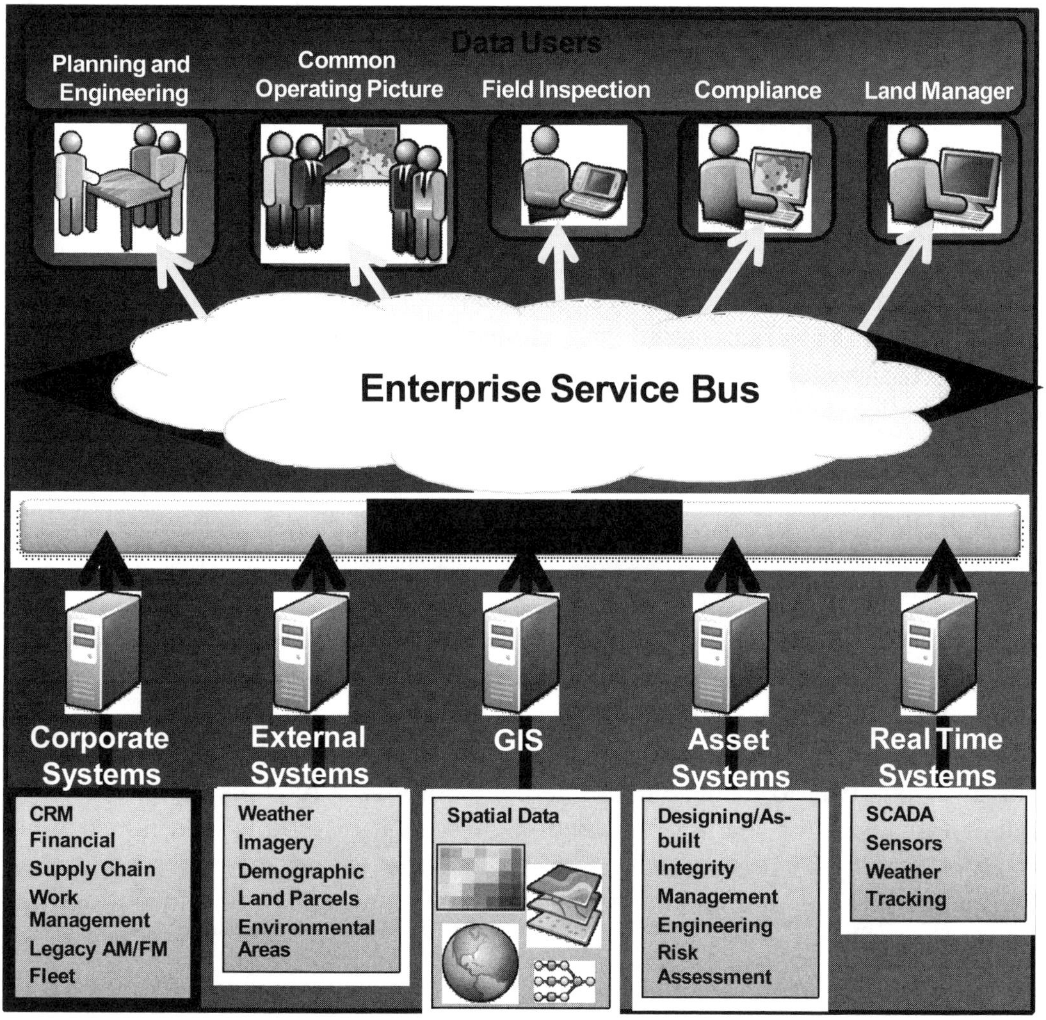

Figure 7 – Example Enterprise Architecture using the Services Oriented Architecture

While direct integration requires tight coupling, the SOA is a loosely coupled system where standardize software functions, or services, are created to be shared by numerous dissimilar applications and technologies. Essentially, the point-to-point integrations are replaced by flexible, reusable integration components that merge the enterprise applications and transactions into enterprise services. These services become vehicles to launch operational processes and tasks from within an external application. An example of a pipeline SOA is illustrated in Figure 7. The figure illustrates how web services are used to bring information from different systems together.

Let's examine this further using an example where a GIS team wants to expose a location tool to other enterprise applications. To start, a GIS administrator has created a Web service that encapsulates a standard piece of functionality—queries street addresses related to an input parameter, such as the names of field offices, for use by other applications. In a pipeline operation there are many applications that might want to know specific addresses such as a customer relationship management solution, fleet management system or an enterprise reporting application. All these applications need to know is how to send the necessary information to the service to get the answer they need. So in SAP, for example, a user has the name of a field office and wants to know the location. The user simply selects the office record and chooses the location button in the SAP interface. This button has been programmed to execute a function that provides the field office name to the GIS Web service. The office name is used to interrogate the spatial data for the address. When the address is located, the data is sent back to SAP and displayed for the user. This process would be the same for any application.

2.6.1 Interoperability

The last topic we will cover in this chapter is data interoperability. Geospatial data comes in hundreds of file formats and is be created and distributed by many organizations worldwide. To ensure that information can work together an operator needs to ensure that data is interoperable. An enterprise integration strategy is one way to manage data interoperability as it requires that a company establish methods for sharing information. A second method is to create spatial, metadata, and content standards. These standards define the types of spatial coordinates systems and datum that information is accepted in, and outline the content for attributes.

Regardless of how many standards exist and how diligently a company polices its data, interoperability issues will arise. It is essential that the company implement a GIS that reads and integrates many different data types, and supplements its capabilities with applications that can reproject spatial data and convert information from one format to another.

2.7 Spatial Data Case Study

To fully understand the value that spatial data has, and to better explain data maintenance, we are going to use an example. In this example a pipeline operator builds a new pipeline to satisfy regional gas demand and then perform some regular activities once the pipeline is in operation. The company has selected an engineering team, and decided to design the pipeline internally then contract out the physical construction. Initially the engineering team accessed some very general mapping data and created a regional map. This provided them with a general lay of the land and was used for initial project meetings.

While the initial map provided some direction to the project, it did not include any of the detail necessary for initial planning. So the team requested GIS data from the local counties and the United States Geological Survey (USGS). They received spatial data layers including surficial geology, soils, vegetation, population centers, roads, land use, and a digital elevation model (DEM). Using a GIS, the team combined the layers to determine several optimal pipeline corridors. They then used the GIS to perform an analysis to determine the best path. This GIS-based model supported the company's desire to minimize construction costs. Requirements included limiting vegetative clearing and elevation changes, using accessible land (agricultural property), avoidance of populated centers, and keeping the pipeline as short as possible. While there were several route options, one was clearly the best choice.

During this project the team made several submissions to external agencies that regulate pipeline industries. One of these submissions required an environmental statement. The team researched and purchased detailed spatial data for vegetation, hydrography, and animal habitat, as well as fine-resolution satellite imagery for the pipeline corridor and surrounding region. Using a GIS extension, the imagery was classified to validate and supplement the purchased vegetation data.

The team then conducted field reconnaissance to confirm all the information and to collect additional environmental information. Field information was recorded using a mobile GIS client and then merged with the existing spatial data. The new environmental information and the data used to select the pipeline route were layered in the GIS and a series of environmental alignment sheets were generated. The maps were used to support the government submission and for town hall meetings with local stakeholders and landowners. In addition to the maps, the team used GIS to create a 3D model of the area to mitigate concerns.

Once the project was approved, the initial pipeline design was completed and migrated into the GIS for use with the existing data. The next step was to acquire land rights. The land department started negotiation with landowners. Using existing data, including county land data, a series of map books was produced for all the land staff. The map books showed regional properties and ownership information. When land negotiations were complete, the new property contract numbers were added to the property data and a final land layer was created for use in the GIS.

To facilitate the construction process, the GIS team created a series of maps and digital data that was used to survey and stake the route, clear the right-of-way, and mark all foreign assets. As construction proceeded, a survey crew used a real time kinematic (RTK) GPS system to record all the construction details—the centerline of the pipeline, welds, points of inflection, inline assets, rights-of-way, and third-party assets such as foreign line crossings. This information was submitted regularly to the project manager for review. When construction was completed, a set of as-built drawings was submitted documenting how the pipeline was constructed. In addition, the transmission company received a database of all the survey information.

The survey database had to be added to the company's corporate pipeline data model. The transmission company used a bulk loading utility to migrate the information from the nonstandard form, preferred by the engineering company, to the standard model that the operator used. During this process, features were moved into the required tables and the supplied codes were transformed into the standard domain values. The information could have been captured from the as-built drawings, but it would have been more time consuming and costly.

When the migration was complete, the integrity of the spatial information was checked by generating a series of alignment sheets for the new line. The sheets contained spatial layers for the pipeline including an aerial photo background, inline assets, foreign line crossings, routing notes, construction notes, topography, and property ownership. Using the GIS, any data anomalies were corrected. Following this, the new pipeline information was made available to the rest of the organization through a series of existing GIS Web-based services. These services were consumed by the supply chain, maintenance, and financial applications. Documents were linked to the new assets using a GIS-based toolbar that searched the document management system using key terms.

Although construction was complete, the team still had many tasks that required the use of spatial data. They needed to create an integrity management plan, perform risk analyses, design a maintenance program, route field staff, perform field data collection, calculate the pipeline class and high consequence areas, and define a public notification program. Emergency response plans were also created using the pipeline database, landowner information, the road network, and regional staffing. These regional plans used the GIS as a data access point for regionalized information on the available staff, their qualifications, and then designated internal and external first responders. Finally, reporting templates were created for yearly regulatory submission.

The pipeline was operating within parameters, but after three years, the transmission company decided to perform an inline inspection. A series of alignment sheets was created and provided to the inspection company. These sheets depicted the pipeline route with an aerial photo background and the relevant inline assets (reducers, mainline valves, traps, launcher, receivers, etc.). The inspection company used the sheets to plan the inspection so that the PIG did not get trapped in the pipeline. A geo-PIG was used, and the results were provided to the transmission company in a spreadsheet. Once an initial investigation of the inspection results was complete, it was discovered that the inspection data did not match the pipeline measurements and the domain codes were different. The information was integrated into the data model using a specialized GIS-based tool that allowed the operator to standardize the feature codes and then find like features in the pipeline database and the inspection spreadsheet so that the information could be spatially aligned. The operator began by looking for mainline values and connecting the data. The process continued until the level of accuracy was acceptable.

Once the data was added to the database, a new set of alignment sheets was created. The inspection alignment sheets were analyzed by an integrity engineer, who determined that one area needed to be physically inspected. He filled out the required forms and, using the GIS as a portal, created an inspection task in the maintenance application.

The maintenance request was relayed to a field inspection team as a task in their e-mail system. The team members needed to locate and mobilize the necessary equipment, so they accessed a GIS layer on fleet and equipment, and scheduled the equipment and the necessary staff. On the specified day, the crew and equipment arrived at the anomaly site where, using a mobile GIS tool, they located the section and exposed the pipeline. While the team performed the physical inspection, they recorded findings with a mobile tool and linked pictures using a Bluetooth camera. The information was analyzed by an engineer and a repair plan was created.

The engineering department sent a field construction crew to the anomaly site. During the repair, GPS was used to record changes to the pipeline and other assets. This survey information was sent to an operations technician in the maintenance department. Using GIS, information was updated in the pipeline data model so that it was kept current.

As this example has shown, GIS and spatial data provide value throughout the pipeline lifecycle from planning and construction through operations and maintenance.

2.8 What is the Future of Spatial Data Management?

This chapter was intended to provide insight into what makes information spatial, and why it is valuable to any pipeline organizations. While we have outlined what can be done using spatial data and GIS, the technology is just beginning to come of age. The immediate future of spatial technology will focus on the application of existing solutions such as integrity management or risk assessment. Where will we go from here?

The future of pipeline GIS is centered on accessibility and ease of use. GIS has traditionally filled a niche within pipeline organizations, used by people who have been trained to operate the software. The future of GIS involves the movement from a niche to the entire organization. There are two ways this can be done.

First, the services oriented architecture information will be provided as a Web service and consumed by other enterprise clients, thus geo-enabling these other enterprise tools. While this type of spatial enabling of business applications has always been possible, it is now simplified and promises to substantially grow the pipeline GIS market.

Second, users will be allowed to access and mash up content with a standard Web browser. As pipeline staff becomes familiar with looking at things through a Web-based geospatial view, they will be increasingly interested in services that go beyond simple maps and images. The Web platform will be used to serve data, analytic models, and metadata to others to use on the Web. This easy-to-consume environment is cost effective for organizations, and drives real value from spatial data investments.

References

(1) Thomas, C. and Ospina, M., "Measuring Up: The Business Case for GIS," ESRI Press, 2004.
(2) Adam, S. and Hlady, J., "Data is an asset that should be managed," Proceedings of the International Pipeline Conference (IPC 2006), Calgary, Canada, ASME, 2006.
(3) A good general resource is http://www.usgs.gov/pubprod/
(4) Refer to http://www.opengeospatial.org/
(5) Refer to http://www.isotc211.org/
(6) Refer to http://www.pods.org/
(7) Refer to http://www.apdm.net/

Acknowledgements

The author of this chapter gratefully acknowledges Jessica Wyland for her efforts and contributions to this chapter.

3 Pipeline Engineering and Construction

3.1 Introduction

The design and building of pipelines has made use of geomatics through traditional field surveys. Land surveyors would select the route and then collect data for alignment sheets or for submitting plans with the land titles registry. Modern geomatics professionals still assist in these tasks but advances have helped pipeline designers and builders deal with more data, address more regulations, and effectively maintain pipeline systems better than they were able to in the past.

This chapter discusses those aspects of pipeline design and construction that provide some context to the reader as to why and how geomatics and specifically Geographic Information Systems (GIS) could affect and interact with pipelines during design and construction. The chapter approaches the design and construction of pipelines, at an introductory level, using the framework of a pipeline project. Then we will touch specifically on areas where different classifications of pipelines require further detail. This will also help with introducing concepts of GIS and where the pipeline industry currently sees the benefits of this discipline.

Like any other business venture, pipeline design and construction is directly driven by the need of a customer, a group of customers or an overall market. Whether it is a cost of service, long-term contract or a toll per volume, the project has to be economically viable. This is highly related to and forms some of the reasoning for using new technologies such as GIS.

In looking at design and construction for a pipeline, the type of construction needs to be taken into consideration, whether concentrated in an area or disposed over a linear distance. That is, in discriminating between construction of line pipe or above ground facilities, such as pump stations, metering terminals, and tank farms. The type of commodity also drives cost, level of standards, and of course, operations. Most of the development in engineering for pipeline construction comes from the hydrocarbon industry. This industry generally involves the use of carbon steel pipelines that have to have different design parameters according to whether the product is gas, high vapor pressure liquids, low vapor pressure liquids or a multiphase flow.

When talking about the hydrocarbon value chain, there are pipelines on the upstream, midstream and downstream segments of the industry. Pipelines in the upstream side of the value chain are generally owned by the oil producers who take the commodities from the well head to the treating facilities for example, to remove sulfur in the case of sour gas, or to remove water in the case of secondary oil production.

After the commodities have been brought up to "sales specification" or to a quality that is generally standard within the midstream industry, the commodity is transported by a gathering system which varies in sizes according to the scale of the market, the reservoirs and the scale of the exploitation in a specific region. In the case of conventional oil pipelines, the gathering point or "custody transfer" point is generally called LACT (Lease Automatic Custody Transfer). For conventional gas pipelines, the custody transfer point is called a "Metering Station", whereas with conventional oil, the gathering systems trend to be an intricate network of many custody transfer points that form relatively large streams. In the case of oil sands production, the gathering systems tend to be large pump stations adjacent to large treating facilities.

The midstream large transportation pipelines generally take product from the gathering pipelines at tank farms where the production has already been collected. From there the product awaits "its turn" to be taken by the trunk line, either as a "batch" or as a "slip stream".

The trunk lines generally deliver product to refineries, NGL plants or marketing hubs where the oil production can be taken by another trunk line and then shipped, either across the continent, locally, or to another continent.

The downstream pipelines would generally be considered the distribution networks. These are used by downstream refineries or NGL extraction plants to transport the "refined" or "straddled" products to distributors or chemical plants.

In all cases, pipelines form the basis of delivering a product over distances. We know that the longer the pipeline has to travel, the greater chance it will come into contact or into close proximity to facilities such as wellsite and refinery, while minimizing contact with population and avoiding complicated crossings. Inherently, GIS is a technology and a set of processes that allow pipeline designers and builders to have a better grasp of issues of proximity, ensuring that a pipeline or facility is designed and built in an economic and responsible manner. The following section builds a framework for a geomatics practice that is becoming a fundamental piece of pipeline projects, one that forms the cornerstone to taking advantage of the volumes of digital data available to designers that is later migrated to the operator; this is data centralization.

3.2 Data Centralization for Pipeline Projects

This topic is thoroughly covered in Chapter 2 under the title of spatial data management. However, this small section summarizes those concepts in a way that they can be applied to design and construction. A fitting framework to pipeline GIS is the concept of the *Central Database*. Since most forms of design, modeling, and analysis require an input of data, the central database provides a robust support to these operational realities. In fact, the central database is as much a philosophy as it is computer infrastructure, software, and processes. By subscribing to the philosophy of *a single central-source of truth*, pipeline projects and owner/operators begin to treat data as a valuable asset, keep data organized, secure, accessible, and effectively managed through continuous changes, updates, and applications. Once data is respected, decisions founded on that data gain value. The term *Paradigm shift* is commonly used to denote a punctuated change in a discipline's thinking. Thomas Kuhn (1) brought this term into its modern usage by referring to scientific revolutions as paradigm shifts "Successive transition from one paradigm to another via revolution is the usual developmental pattern of mature science."

The central database philosophy fits into this model since it is causing the engineering disciplines to approach problems in a *data-centric* manner, one that is fundamentally new to the discipline. Furthermore, the central database approach has a sense of inevitability since the amount of digital data accessible and the volume subsequently generated will only keep growing. As we will see throughout this chapter, data is the basis of many automated processes and the central database philosophy is the foundation to fully leveraging data's use to support pipeline engineering.

The main reason for having a central database is to manage change. This is achieved by providing an information infrastructure for a pipeline's lifecycle and adds stability to an information management system. The structure it provides allows data users to conduct their activities in a stable environment despite the constant changes being made to the data. With a dependable information system backbone, data use and sharing become valuable additions to a project (2, 3). A central database provides this stability through four supports to the information management system. These are:

- Direct change management
- Data security
- Data integrity
- Metadata (the data storyboard and legacy)

These four functions of a central database allow for the centralized control and quality assurance of all data. As a result, an organization will have a single source of truth despite the variety of disciplines and contractors contributing data and generating information. Through planning and construction, operations and integrity maintenance, and finally to decommissioning, the central database's contents and physical infrastructure may change, but the four functions will continue, thus providing the desired stability for the life of the project or enterprise.

3.3 Front End Engineering Design – Line Pipe

A GIS is a tool that allows users to model or simulate the real world by introducing and manipulating controlled variables, and analyzing the output. The purpose of building and using GIS models is to leverage information from data by replicating or estimating a real world phenomenon. The resultant information can be analyzed and used to make better decisions during pipeline system design or operations.

Models within a GIS are essentially mathematical equations with real world spatial references that attempt to predict real world situations. By modeling scenarios in a GIS, complicated processes are automated. These automated processes can be repeatedly tested whereby, expensive field work is minimized, cost estimates can be refined, and project scopes narrowed to increase overall work and cost efficiencies. During this stage, the line pipe design can be an iterative exercise that is highly related to the producer or customer's volumes available to be transported. The physical properties of the product to move, and to some extent, the chemical properties of the product, also play a key role in the sizing exercise.

Modeling can be used in all stages of pipeline system design from pre-field work planning, detailed design, construction, and operations. By modeling on the desktop, engineers and project managers can become familiar with the project area and plan for the terrain, project variables or other phenomena that may affect their pipeline projects. With the ability to model design options from the office, projects can be safer, costs can be reduced, and scopes refined. Some common GIS engineering models are shown in Table 1.

TABLE 1 – COMMON GIS-BASED ENGINEERING MODELS

GIS Model	Applied Uses
Buoyancy Control	Using terrain type modeling to determine the amount and type of buoyancy control to use in areas where the pipeline traverses swampy terrain
Cut / Fill / Volume	Using a DEM (digital elevation model) to calculate how much earth to move, remove or add.
Least Cost Path	Manipulating multiple variables to determine the best routing for a pipeline
Terrain Analysis	Determining various geohazards, such as slope creep, frost heaves, or acid rock drainage
Line of Site	Determining how visible a pipeline will be from a town or designated point of view
Cross and	Calculating slope and its affect on pipeline routing (used

GIS Model	Applied Uses
Longitudinal Slopes	with or without other variables)
Hydraulics	Using a DEM to calculate the hydraulic flow rates of a pipeline system and determining the optimal location for pump or compressor stations
Hydro Networks	Compiling a stream network, to model a change in stream flows resulting from the influence of a pipeline and its effects on the local environment
Pipeline Crossings	Analyzing various ROWs to determine optimum pipeline crossing locations
Stationing (Linear Referencing)	Using DEM's to calculate chainage of pipeline systems and features along the ROW.
Air Quality Monitoring	Determining general wind directions and their potential effects on the spread of exhaust from facilities
Dynamic Segmentation	Transforming tabular data into segmented linear features using from and to measures
Timber Volumes	Calculating the volume of trees cleared during pipeline construction
Risk Models	Estimating external corrosion, internal corrosion, and third-party damage

Models cannot provide all the answers for pipeline design and will not replace field work entirely, but they are valuable tools that aid the decision-making process. The value in using GIS models is that they empower engineers and designers to simulate countless scenarios and sensitivities, thus enabling more informed decisions.

A good example of GIS-based design modeling is "running hydraulics". In order to run preliminary hydraulics, we obtain an elevation profile. At the early stages, a tentative route selection can be made as a "desk top" exercise by using government sourced digital data which may help in determining a preliminary route. The route selection process will be refined throughout the development of the project and can involve field visits and extensive use of available geographical information in a routing model.

During the preliminary hydraulics analysis phases, there is a constant interaction between the engineering and the commercial teams as the proposed pipeline's characteristics are fine-tuned to a project that is economically viable for the customer, and executable by the company that is building the project.

The results of the hydraulics process produce different cases at a very high level. These different cases outline the respective pipeline lengths, diameters, proposed locations for pump/compressor stations, required operating pressures and specific horsepower required for different types of products and volumes. Figure 1 shows some route alternatives overlaid onto coarse resolution government sourced topographic data, which is usually free of charge and is geographically extensive enough to be available for any project.

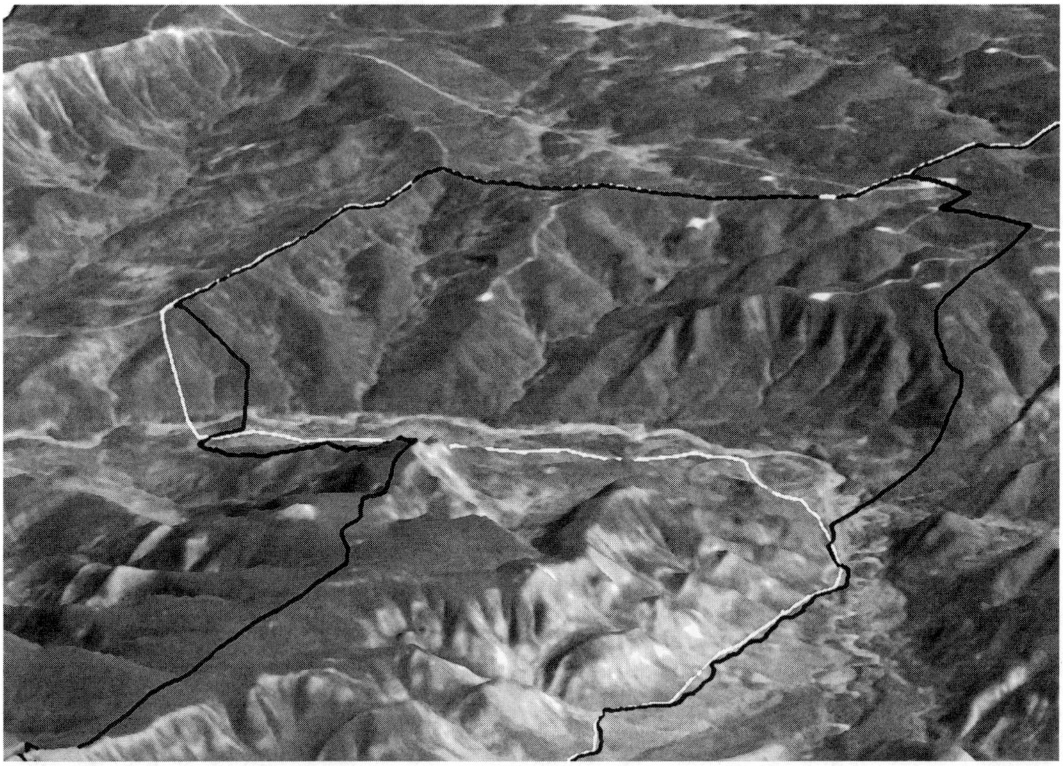

Figure 1. Routing options through rugged terrain based on hydraulic and other analyses

With this information it is possible to establish high level cost estimates and construction schedules, both of which are impacted directly by variables that can be identified by the use of GIS. Common variables are; the amount and nature of river crossings as well as the number of other crossings such as utility power lines, utility gas, roads, government road allowances, railways, residential versus rural areas, characteristics of the terrain, type of soils and soil conductivity. All of these help to determine such things as whether the construction is on dry prairie sandy soils or on wet muskeg that lies under frozen areas.

A draft of the construction and operating philosophy needs to be in place in order to choose between high yield strength and low wall thickness or low yield strength and greater wall thickness. Both have advantages and disadvantages for construction and operating conditions. For example, higher yield strength and lower wall thickness helps in reducing the amount steel to be put in the ground, reducing the purchase, transportation and perhaps construction costs. High yield strength can be associated with brittleness and faster crack propagation on stress corrosion cracking failures on gas pipelines. One advantage with the use of higher wall thickness pipe is having the option of using higher corrosion allowances. With greater wall thicknesses it takes more time for internal corrosion failures to result in leaks during the operation of liquid pipelines, however, heavier wall pipe takes more effort to procure and install.

3.4 Front End Engineering - Facilities

Because of the required transportation distances, product qualities, elevation profiles and amount of volumes to be moved, it is common that the initiating pump/compressor station isn't able to move the product all the way to the desired final destination without installing excessively large diameter pipelines. This creates the need to install intermediate pump/compressor stations. Pipeline geomatics can play a crucial role in determining the optimum location for the station. By optimum we mean that the access for construction and maintenance should be appropriate and cost effective and the terrain has to be taken in account to prevent future flooding and exposure to the elements. At the same time, the location needs to consider mitigation of potential catastrophic failures as a result of leaks, in the case of liquids pipelines, or explosions, in the case of gas pipelines. Ultimately, all of these considerations must pass through public consultations and regulatory scrutiny as well. Facility siting is an issue where the hydraulic engineers and the geomatics expert need to work together for the most cost-effective alternative in the interest of smooth construction and operation of the station.

The design and placement of facilities is always greatly assisted with visualization products (4, 5). Using techniques such as 3D surface renderings, slope maps, and imagery engineers can evaluate topography and physiography where pump stations, staging, waste, and camp areas can be optimally located with minimal 'cut and fill' requirements and thus minimal construction and logistic costs. Geological concerns, such as landslide or avalanche areas, rock deformations, water drainages or land instability can be easily identified, and considered for optimal and cost-effective pipeline construction methods.

Part of the reason for running hydraulics models is to estimate the location of the pump stations. The amount of required horsepower for the pump station is a determining factor for its design size and construction. This compounds itself throughout the size of valves, pumps, piping, cables, motors, transformers, starters, VFDs (Variable Frequency Drives for variable speed on the motors), switch gears, main breakers, and even to the utility service.

Figure 2. A 3D rendering of a tank farm and terminus for both design and public consultation purposes

In the last few years, when a rapid economic growth has taken place, it has been challenging for the existing infrastructure to keep up with the industrial growth, especially for pipelines in remote areas where the electricity transmission and distribution networks may require upgrades in order to provide the necessary power required. Planning for upgrades on the electricity network may need to be kicked off at the same time as the pipeline is in the FEED (Front End Engineering Design) phases in order to avoid expensive start-up delays because of unavailability of electric power to run the stations. This is naturally aided by GIS since it helps designers with proximity issues. When utility information becomes part of the project database, the project team can determine and design for power issues initially from the desktop and position facilities accordingly. In addition, solid change management procedures for facility site scenarios will help the design team stay organized as they change the location of facilities.

3.5 Detailed Engineering

3.5.1 Overall System – Detailed Engineering

At this stage the commercial negotiations with the customer would be almost complete. The negotiations cover off on items such as; the type of contract in place and how the cost structure will work. Furthermore, if a project had the foresight to start data management early, much of the cost of data acquisition and system set-up may be complete at his point.

Within the context of detailed engineering of the overall system, traditional methods of designing, selecting, and maintaining a pipeline route involve manual drafting; a very time and labor intensive process. Increasingly, GIS is being recognized as a valuable tool that can increase efficiencies, while also maintaining high quality cartographic outputs. During detailed engineering, GIS continues to be used to create and maintain spatial data, design and evaluate potential route locations, and perform analysis or modeling to determine potential impacts, as well as produce high quality cartographic outputs in a variety of formats.

All of the tasks below are essentially engineering procedures that produce data for the central database and/or use existing data for their development. The main advantages of collecting this data as we progress through the project are that data have integrity and any change can be managed. We never engineer anything perfectly the first time, hence change management is critical.

3.5.1.1 Definitive hydraulic models and topographic influences

This requires precise topographic information compiled from stereo air photos or LiDAR data (Refer to Chapter 8 for a full description), because the topographic model should be as fine as 1m (3 foot) contours for detailed engineering. It is at this stage that the most intensive use of geomatics and GIS analysis usually takes place. Doing so provides the opportunity to use the most up to date technology in favor of making concise routing decisions that will positively impact the construction and operation of the system. Along with a definitive volume forecast and known commodity properties, either for gas or oil, the findings from the GIS analysis help form this detailed process.

3.5.1.2 Temperatures

The operating temperatures at receipt points where the product enters the pipeline system have a bearing on the type of coating selected. However, constraints on the maximum temperature for a specific coating may require the producer or owner of the commodity to cool the product before the pipeline takes custody.

With underground pipelines that run for hundreds of kilometers while taking the product to its final destination, the product temperature tends to equalize with the ground temperatures along the pipeline. This is why the winter and summer ground temperatures play an important role in estimating the capacity of the pipeline or in sizing it in order to achieve the desired flows. The type of soil and its conductivity also play a role in how fast the product cools off in the first couple of kilometers, and how much heat it will lose along the pipeline. GIS and databases containing soil information for areas where the pipeline is intended to be built, help with providing factual data affecting the pipeline capacity. This approach is far more accurate than using assumptions, which can be common, even during the phases where detailed engineering takes place. Since ground thermal properties change with geography, defining these early on will help fine routing decisions made throughout the design.

3.5.2 Line Pipe – Detailed Engineering

3.5.2.1 Pipe wall thickness, Hoop stress calculations, Selection of materials and Yield strength

This can be an iterative process where, for a specified pressure a wall thickness is calculated based on the assumption of having a specific material with particular yield strength. Alternatively, once the required wall thickness is established, yield strength is calculated based on the application. This process is critical in the sense that it will establish a number of design elements. These include the type of production welding to be used during the project, preheating requirements for low temperature days, depth of cover in the case of underground pipelines, and the weight of each pipe run. Moreover, costs related to procurement costs, installation, inspection and operation can also be determined. The pipeline design information is contained within the CSA Z662-07 standard, clause 4.3.

Regardless of the outcome, this information is stored in the project GIS database and updated if changes are made during design or in the field during construction. This information can then be handed over to the client where they can use it in their integrity program.

3.5.2.2 Pipeline coatings

A common coating used in conventional pipelines has been the Yellow Jacket. The purpose of this coating is to mitigate external corrosion on the pipeline, as well as provide some mechanical protection from rocks during installation that could otherwise scratch the pipeline. Since the pipeline has to be welded in the field prior to laying it down in the ditch, coating is also required at the welded joints which would otherwise remain exposed. These coatings are called sleeves, and are most often installed manually after welding of the joints has been completed.

As technology advances new materials are continuously being developed. This provides the opportunity for new types of coating materials to be introduced to pipeline construction. One such material is the FBE or fusion bond epoxy. FBE is now considered to be one of the best technologies available for external coatings and its use is becoming increasingly common on new projects.

Like wall thickness, coatings information is retained in the project database and incorporated into alignment sheets and passed onto the client.

3.5.2.3 Depth of cover

The depth of cover is related to the amount of soil between the top of the pipeline and ground level. This is very important because there needs to be an appropriate balance between a safe depth of cover that accounts for variations during construction and the cost of digging the trenches. Furthermore, depth of cover affects the type of machinery, industrial safety during construction, and even temporary soil storage between the excavation phase and backfilling activities. Furthermore, appropriate burial depth will take account of the type of soil drainage and farm machinery being used in the vicinity. For example, CSA Z662 has requirements for a minimum depth of cover where pipelines traverse flat terrains, or cross other features like roads and railroads. This can be calculated well in advance of construction using GIS project data that would show crossings along the route as well as topography.

Depth of cover is a valuable piece of data that should be recorded and maintained in the central database since it can be used during operations. Any future geo-pig run will show the position of the center line of pipe, therefore, by knowing the depth of cover, engineers cans see if there is pipe upheaval or other movement.

3.5.2.4 Class location

At a high level, the safety margins to be used during design also depend on the likelihood and severity of accidents in the context of potential human loss. That is, a pipeline that runs through a rural unpopulated area has different safety factors incorporated throughout its design when compared with a pipeline that runs through an area that is likely to be populated a decade after its installation. These different factors are primarily focused on pipe wall thickness, depth of cover, and potential operating pressure. The class locations, definitions and design guidelines are in the CSA Z662 standard as outlined in clause 4.3.

It is not feasible to determine the number of tenants in all buildings by site visits. Therefore, developers of pipeline standards have predefined the number of occupants based on the structure type. A residential building of a certain size or height has an assumed population density. Other structures have similar predefined population densities. Based on the population and structures definition provided by the relevant codes or standards, a map showing building structures and their proximity to the pipeline allows for creation of a high consequence area (HCA).

Aerial and satellite imagery have been very effective datasets for establishing and updating structure identification and location. (6,7). When there are changes to the pipeline or surroundings, new maps of structures and pipeline are input to the GIS to automatically recalculate a new HCA. Chapter 4 deals with this topic in detail.

3.5.2.5 Acquiring land vs. leasing

In the past, common practice was to establish leases with land owners. However, it is sometimes necessary to acquire land for ROW (Right of Way) in order to facilitate access for construction, or to facilitate reliable operation of the pipeline including; repair digs, access for cathodic protection surveys, access to mainline valves, etc.

Land information systems built upon GIS have been very effective in keeping land owner information and survey plans (8) and even information on easement payments and agreements. Cadastral information is the basis and essentially the origins of geospatial data and is easily handled and maintained in a GIS.

For example, a land system can contain corporate data (e.g. were their pipeline and assets are) as well as have a link to the land titles office to maintain up to date landowner information.

3.5.2.6 Pipeline routing and ROW selection

Another area where geomatics plays a key role is routing in preparation for the construction phase. In fact, this has been an area of intense activity and is an early computational innovation geomatics has provided pipelining (9,10). Factors such as nearby highways, parks, pipelines, land owner support/rejection, type of land use and potential new business can determine the routing of a project. The geotechnical analyses and surveys along the proposed pipeline ROW reveal crucial information regarding the amount of rock, swamp or sandy soil that the construction crews would face. Terrain consisting of rock formations as opposed to muskeg has very different excavation requirements that are influenced not only by the equipment type, but also by the logistics of getting materials and construction crews in and out of the job site. Another description of routing from a geotechnical point of view is found in a recent ASME book addressing Pipeline Geo-Environmental Design and Geohazard Management (11).

Geotechnical information from a GIS can help determine whether or not a project is even viable. Alternatively, competitive advantages like local sources of construction aggregate can be determined using GIS. Issues with environmentally sensitive areas can also be identified at an early stage through the use of GIS. Other entities or variables where GIS provides insight include; wildlife migration routes, fishing and spawning zones, archeological protected areas, previously contaminated soils, irrigation districts and first nation reserves etc. These variables all have different cost implications. GIS is a valuable tool that can help with quantifying these cost implications in an effort to obtain the least cost path, or the most cost-efficient path.

The objective of GIS facilitated Pipeline Routing and least cost path analysis are to minimize capital costs by minimizing field visits, survey costs, operational, and maintenance costs and to address environmental, socio-economic and cultural impact of the pipeline. In today's world of desktop routing, the use of the digital data such as satellite imagery, LiDAR/contours, and other overlays of interest, allows pipeline engineers and geoscientists to visualize their environment in 2-D or 3-D as often and as long as they need to make design decisions. It is important to understand that the model is influenced by the inputs of the users. Therefore, users are required to provide the desired business rules and parameters to answer the two main routing questions:

- What must we avoid (and by how much)?
- What should we be close to?

In order to run different routing scenarios criteria and data must be compiled to feed the rules for a least cost path. The following are some typical data sets for a least cost path routine.

TABLE 2. TYPICAL ENGINEERING CRITERIA FOR ROUTE SELECTION

Engineering Criteria	Data and Information Required
Start and end points	Facilities, accessibility, power
Slopes (cross and long)	Topographic data (DEM or contours)
Crossings	Other pipelines, roads, hydrography, other features visible on imagery
Trenchability	Surficial material
ROW preparation	Surficial material, topographic data, imagery
Buoyancy control	Surficial material, topographic data, imagery
Pipeline facilities	Topographic data, accessibility, cities
Sensitive areas	Protected areas, wildlife, land cover types,
Cultural areas	Cities, land use (e.g. farming)
Land ownership	Cadastral

Computing a least cost path is simple; the desired criteria are selected, criteria is matched with data, criteria are weighted, and input in to the model. However, it is the criteria, or model rules, that are always the challenge. Therefore, many scenarios are typically run so the users can gain an appreciation for the sensitivity of the model to the parameters. A general workflow for a least cost path could looks something like Figure 3.

Given the advances in computing power and growing volume of digital data products, desktop routing presents numerous benefits, including:

- Run multiple options quickly once base data and cost path criteria are set-up
- Easy to implement engineering changes (ie. new data, new start / end points, different parameters: slope values, distance from existing features, etc.)
- Absolute or relative costs can be part of the routing algorithm in order to objectively evaluate the difference between routes
- Office time vs. Field time: safety and time cost savings
- Can be done in different geographical locations (example: can be run from Houston office for a project in Alaska)
- Can be done in-house
- Creates a standardized process and maintains business rules some knowledge is retained
- Facilitate the multidisciplinary effort and input (Environmental, Geotechnical, Engineering etc.)
- Facilitate different political , social and economical requirements pertinent to different project area
- Applicable at all stages of engineering, only the granularity of the input data must change.
- Place a standard for route selection as a starting point for all projects

3.5.2.7 Regulatory and development permits

In the same way that some of the detailed engineering deliverables have to start early in the project, the regulatory process and clearing of any potential hearings or public consultation has to happen well in advance of other activities. The viability of a project may depend to a large extent on the regulatory permits. Even if a project is viable from the perspective of hearings and public consultation, the schedule will be dependent on the timing of the various regulatory permits. The regulatory process will hold a very high priority within the critical path of the project. This component is covered in detail in Chapter 6.

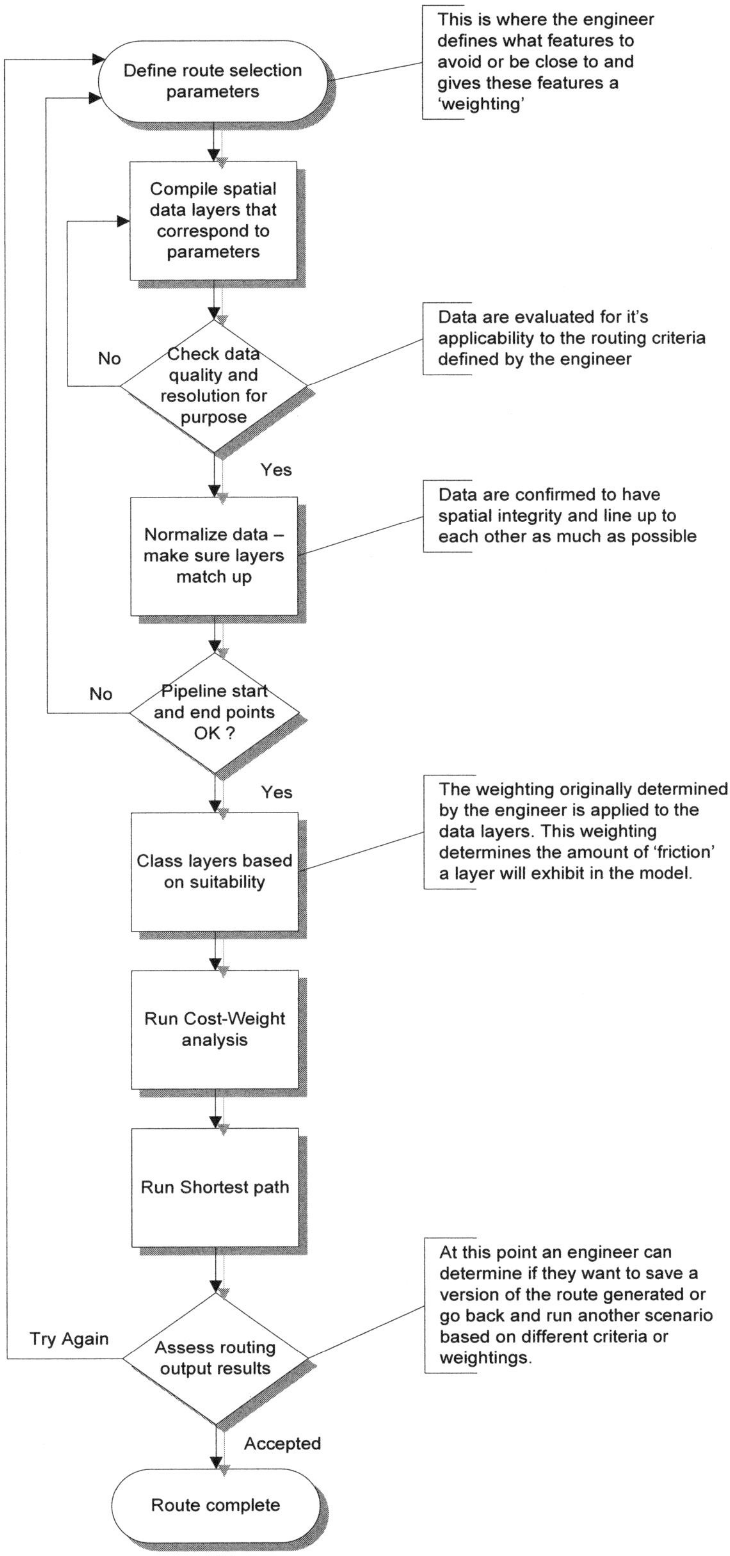

Figure 3. General workflow for least-cost-path route selection

3.5.2.8 Creation of Alignment Sheets issued for Construction.

Automated Alignment Sheets play an important role in pipeline projects. They provide a visual representation of where the pipeline exists in the real world. Over time, alignment sheets have become richer in content as larger amounts of digital data are captured to aid in design. The use of Computer Aided Drawing (CAD) has significantly aided in the creation of alignment sheets but with pipeline projects becoming larger and more complex, traditional methods of CAD drafting do not suffice in managing the increased digital data.

With the use of databases and constant improvements in computing hardware, developers have been able to create programs that handle the more tedious areas of alignment sheet generation, namely the data management process and sheet generation process. Automating the creation of alignment sheets does not result in products that are radically different from those produced manually, but rather the new technology results in time savings and reductions in overall effort.

Alignment sheets are the main deliverable from the detailed design process as the project moves into its next phase which is construction. The alignment sheets typically display a variety of environmental characteristics related to the area that the pipeline is going through. Common examples include; swamps, river crossings, utility crossings and road crossings. Depth of cover in the area, wall thickness variations as required by the different crossings and class locations for the pipeline would also be represented. Even detailed engineering specifications like pipeline MOP (maximum operating pressure), material grade, mainline valves, risers, loop connections, pig traps and the proximity of other pipelines are included.

3.5.2.9 Traditional Alignment Sheets

The traditional method for creating alignment sheets typically requires a large number of CAD operators, with each operator maintaining a specific number of sheets for a designated section of the pipeline. These sheets are updated and changed throughout the duration of the pipeline project and with each manual change, the integrity and quality of the data shown on the sheets is reduced. This reduction in quality is caused because changes made in one sheet are not automatically reflected in other sheets; the CAD operator must manually change them. Even with a proficient and effective group of CAD operators and engineers, the job of manually maintaining data on the alignment sheets ultimately fails as the volume of the data becomes too great or edits happen over time. The use of automated alignment sheets minimizes, if not eliminates, the pitfalls of manually drafted CAD alignment sheets.

3.5.2.10 Benefits of Automated Alignment Sheets

There are definite benefits to using automated alignment sheets. The actual sheet production process takes only a fraction of time compared to manual drafting, and efficiencies are continually increasing with advancements in hardware and software development kits. A single person can easily manage the entire process of automated sheet generation.

In addition to reducing generation time, automated alignment sheets can be very flexible in terms of the data and content included. Specific sets can be created for engineers who need preliminary routing information, and a completely different set can be created for regulators and those government agencies who are interested in environmental aspects of the routing. All this philosophy is consistent with the central database theme where a main repository of data and information reside which is called upon to provide the current version of a dataset for aq particular query or function.

3.5.2.11 Implementation

Creation of automated alignment sheets can only occur if a pipeline project has implemented a centralized spatial database. This is important as the automated alignment sheets use the centralized spatial database as its single source of truth from which all data is extracted. On a basic level, automated alignment sheets are just a complex database report. Each set of alignment sheets is essentially a snap shot of the data within the centralized spatial database.

A shift in thinking is required to ensure the successful implementation of automated alignment sheets. Using the manual technique requires the majority of effort be spent maintaining the data on each individual sheet. If the data is found to be incorrect, the CAD operator must manually correct it. Using the automated approach, this effort must be shifted from sheet-by-sheet maintenance to maintaining the datasets within the centralized spatial database. When a data change is required on an automatically generated alignment sheet, the change must occur in the database and not on the sheet itself. By managing change within the database, all subsequent sheets generated will reflect the most current and accurate data. As a result of this mentality shift, the front-end effort required prior to sheet generation is very important, since each dataset should be verified for accuracy before the sheets are run.

3.5.2.12 Data Quality

Users planning to implement an automated alignment sheet solution must understand that the front-end effort of performing quality assessments and checks prior to sheet generation is vital to the quality of the output product. Automated alignment sheets can be considered as a complex database reporting tool that displays data exactly as it exists within the centralized spatial database. If the data within the database is incorrect, out of date, or incomplete at the time the sheets are run, these data deficiencies will be reflected in the output sheets.

The responsibility regarding data quality rests between the database administrators and the pipeline designers. These two groups must work together to ensure that the data created by the pipeline designers is properly represented in the centralized spatial database and thoroughly checked by the quality control measures used by the database administrators. The channels of communication and responsibilities between these two groups must be well established and clearly defined in the change management process.

In most projects, a pipeline data model has already been accepted and all the input data is parsed into the corresponding tables within it. Whether or not a project has opted to use a pipeline data model, the data within the centralized spatial database must be organized in a manner where linear referencing can be applied.

3.5.2.13 Linear Referencing

Automated alignment sheets rely heavily on a method called linear referencing that is used to integrate the data in the centralized spatial database. Linear referencing is a straightforward method for associating attributes or events to locations or segments of a linear feature. Using linear referencing, events are easily located based on their distance from an established starting point, rather than from an exact *x-y* co-ordinate using measurements of latitude and longitude. For pipeline projects, the pipeline route would be classified as the linear feature and all corresponding events and elements would be referenced to the route via the measure (e.g. milepost). The value of linear referencing is that it *dynamically* adjusts event locations along the pipeline as the route changes, or as alternate design options are explored.

Linear referencing can easily be applied to features such as valve locations, road crossings, creek crossings and utility crossings because these features can only occur at a specific measure along the route. This also applies to features such as pipe protection and erosion control that are required to run along the lengths of the pipeline and are denoted by a start measure and end measure.

Features that are area-based (legal land boundaries, soil types and slope boundaries) can be linearly referenced to the pipeline by the intersection points where the pipeline enters and leaves the area. The success of linear referencing relies heavily on the quality and maintenance of the centralized spatial database.

3.5.2.14 Software Solutions

There are several software companies that provide automatic alignment sheet generation (ASG) programs, but the workflow dictating how sheets are generated is common to all at a high level. The ASG first queries the database and retrieves the corresponding spatial data; then translates the spatial data from real-world coordinates to paper-space coordinates; and, finally generates the sheets according to the layout template specified by the designer. There is very little human interaction between the first step of querying the database to the final step of sheet generation. As a result, the sets of alignment sheets generated are consistent as each sheet was created using the same method and data. This eliminates the potential non-systematic errors that are typically encountered during the manual drafting of alignment sheets.

3.5.2.15 Process Overview

It is important to note that despite the changes in implementation and mind set, the process of generating automated alignment sheets remains iterative. A set of sheets will go through several revisions before reaching its final form. Feedback should always be solicited from the appropriate stakeholders, and a final quality assurance review should be the final stage of the process. Figure 4 illustrates the high level workflows required to generate automated alignment sheets.

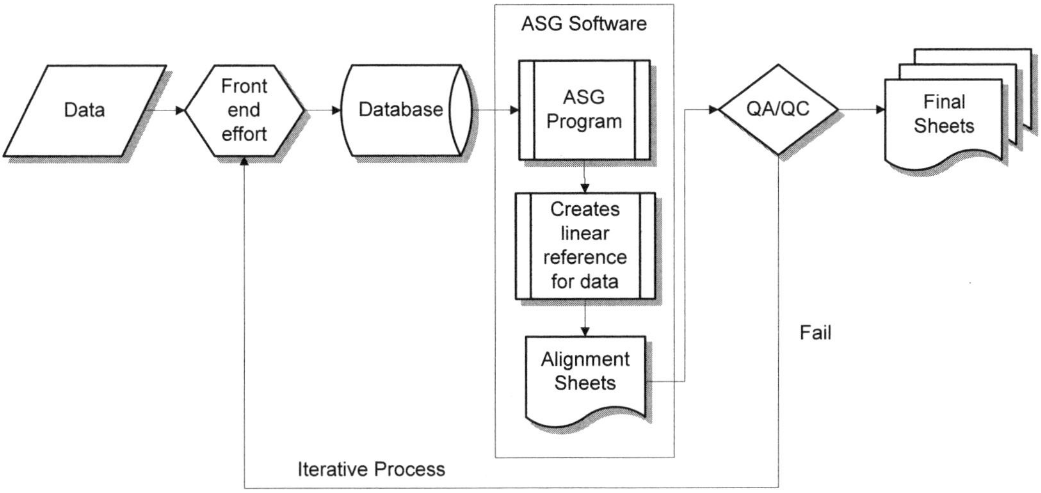

Figure 4. Automated Alignment Sheet Process Overview

The iterative process starts at the "Front-End Effort" and ends at "QA/QC". The process is complete when the sheets have met the specified requirements of the project. Though there are six steps listed in the iterative process, only three require human input. The remaining three steps represent areas where manual drafting would be required in the traditional sheet generation process.

As is evident, alignment sheets are a valuable part of the process and contain vast amounts of information. Using a GIS is an efficient way to manage all of this information and mash it all together to produce an alignment sheet that is detailed and organized.

3.5.3 Line Pipe – Construction

3.5.3.1 Construction ROW, Temporary Permits and Arrangements

Prior to construction, and even during the stages of detailed engineering, vegetation removal or preparation of the ROW needs to take place according to the project schedule. This activity can also provide a reduction of the construction risk by eliminating unknowns and providing first hand information to the detailed engineering team. At the very least, it provides an opportunity for the construction group to understand particulars of the ROW that may have not been identified in the prior stages. In some jurisdictions, it is becoming acceptable to issue requests for permits using GIS data such a LiDAR and high resolution images in place of a field survey (12,13). This has arisen due to the fact that many companies already posses the geo-spatial data required to file an application and can do so without heading back into the field. Therefore, regulatory bodies are looking at accepting these applications as long as the data meet certain accuracy standards.

The right of way used during construction will be wider than the permanent one that will be maintained during operation. The reason for this is that construction requires the transit of large equipment such as excavators, hoes, trenchers, side booms, flat beds, cranes, and even the labor force. The backfilling material and the top soil removed during trenching will likely be stored in two different stacks while the pipeline is being lowered, thus requiring extra space during construction. All this information can be figured out and maintained using the project GIS and route maps or alignment sheets can be produced as needed for field personnel to see.

Depending on the land agreements, the permanent ROW can be owned by the pipeline operator, and the extra room for the construction may be used under temporary permits or leases. In Canada, in the case of Crown land (land owned by the government), leases with exclusive surface use rights may be obtained. In the case of prairie or farm land, there may be a temporary access for construction, but after completion the ROW is usually expected to be left in such condition that it can be used again as farmland. Typically, after a couple of years of crops have been seeded over and then harvested, it is impossible to know that a pipeline is in the ground, except for the presence of risers or mainline valves. Essentially, the ROW land is returned to continuous farming. Going forward, notifications and permissions with potential indemnification for crop damage are required from land owners every time that access is required to this farmland in order to perform a pipeline repair, cut out, or for example to install above ground markers for pipeline inspection jobs. Again, GIS is extremely effective in integrating these different layers of information to support business decisions. At the end of the project, all this data can then be migrated to the client for their operation.

3.5.3.2 Trenching

Once the right of way is clear of vegetation, the permits are all in place, the alignment sheets are complete and issued for construction, and the equipment is in place with an adequate construction team, then it is time to start the excavation. Pipelines built in areas like the prairies and mountains are usually built during the summer months. Swamps and muskeg are usually constructed during the winter months when the ground is frozen. This allows heavy equipment to travel across the surface with a much lower environmental impact. The trencher stock piles the soil at a safe distance so that there is sufficient room alongside the trench for equipment and crews to move back and forth while welding of the individual pipe segments takes place.

Once all of the pipe segments have been welded together into a continuous pipeline, it can be lowered into the trench. The nature of some pipe joints requires them to be welded down in the trench after lowering of the pipeline has taken place. For safety reasons, this requires special grading of the trench sides. The slope of the trench can be 1.5:1 or 1:1 depending on the type of soil and depth of the trench. Typically safety guidelines are available to govern this type of grading. For example, in Alberta, Canada "A Guide to the OHSA Excavations Standard" prepared by OH&S (Occupational Health and Safety) contains the guidelines for this type of work (14).

Trenchability is a model that can be done very effectively in a GIS. This allows a team to assess the trenchability of a certain segment of the route, quantities of excavation, and ditch volume. This model based on surficial material, route alignment, and topography.

In addition, trenchers have recently been equipped with GPS control systems to navigate the pipeline route and accurately excavate. Excavation width and depth can be recorded as well as the precise position of the trench all of which is within inches. Another advantage of this technology is that the route does not have to be re-staked whenever changes are made. The route is loaded into the control system and is used to navigate the equipment; a survey crew does not have to stake out the center line.

3.5.3.3 Pipe Mills and Coating Logistics

The specific pipe identified during the detailed engineering stage will evolve in a procurement order. The pipe is usually sourced from a local mill if available; otherwise it has to be brought in from elsewhere, sometimes even another country. Coating information of individual pipe joints can be stored in the central database so that the operator has a record for integrity monitoring.

3.5.3.4 Bending Pipe

It is necessary to accommodate changes in elevation along the pipeline route from either trenching or the terrain itself by bending pipe. Often a bend of only a few degrees is required as it is imperative to avoid flattening or humps. For this reason, the bending angles are kept quite small wherever possible. Occasionally, pipe bends are required to be at 90 deg, when going around a corner for example. In this case, the radius has to be large enough to be able to accommodate ILI (In Line Inspection) tools during commissioning and pipeline maintenance. Another example is "S" bends for risers, which are documented in the alignment sheets for some pipelines. This information can also be documented in "as built" packages for risers. Estimating with accuracy the number of bends prior to construction can be challenging. GIS provides a ready tool that can be used to accurately tackle this challenge as elevation profiles and planned ROW routes can be combined and examined within a single interface or application.

3.5.3.5 Welding and NDT (Non-Destructive Testing)

In the US, welding is done in accordance with the API 1104, ASME's B31.4 or B31.8 Code. In Canada, pipe welding falls under the CSA Z662 Standard. Most of the welding regulations under CSA Z662 are consistent with the US codes. As mentioned earlier, the CSA Z662 has jurisdiction on the line pipe construction whereas ASME B31.3, the process piping standard is used in facilities construction.

After each weld is completed, Radiography or ultrasonics is used as a non-destructive testing method to ensure that the integrity of the weld, so preventing a future failure incident from this cause.

In warmer southern climates, low temperature material requirements and installation methods are typically not required. In the northern areas, where installation temperatures can be close to -22 Fahrenheit (-30 degrees Celsius), it is common to design for low temperatures. The procedures for low temperature welding involve pre-heating of the metal in order to minimize the presence of stress when operating at much higher temperatures than that present during installation. A very useful piece of information that can be stored in the GIS and issued in the as-built package alignment sheets is the ambient temperature at the time the field joining was performed. This information could be valuable during operation when maintenance welds need to be done, or when cut outs are required for repairs or tie-ins.

3.5.3.6 Commissioning

Usually for line pipe, after hydrotesting, a caliper tool is run through internally to detect construction defects such as dents, and perhaps other debris that may have been left from construction. This information can be used to document the need for future testing, monitoring, defect assessments, or even performing a cutout to remove a defect that could prevent maintenance pigging from being performed on a regular basis. This information is stored in the central database in a similar fashion as the other integrity data that will be stored as the pipeline goes into operation.

3.5.4 Facilities – Detailed Engineering

The development of the detailed engineering for facilities is the next step after having understood all of the requirements related to Process Engineering. The primary facilities related to a pipeline are pump stations, metering terminals and tank farms. The detailed engineering for facilities uses the inputs from the front end Engineering discussed above for line pipe and facilities. These inputs are also used along with the above mentioned codes and standards, when considering the detailed engineering for the overall system such as product properties, process conditions and the development of hydraulic models.

Process Flow Diagrams (PFD's), Piping and Instrumentation Diagrams (P&ID's) and the hydraulic analyses are the main documents from which the majority of the detailed engineering originates. Having obtained the required power from the detailed hydraulic analyses, this helps complete the information required by the P&ID's. A control philosophy narrative is developed in conjunction with the P&ID's, in order to outline how the equipment is expected to operate. This is inextricably related to the actual business arrangements performed during the conceptual development of the project. An example of this would include determining who controls the flows into or out of the pipeline system. The contract will also detail the implications of receiving a product whose quality is outside the expected parameters as specified within the contract.

3.5.4.1 Locating Pipeline Facilities

The siting exercise utilizes a combination of information to determine the most optimum location of a facility. Information from a hydraulics stand point, access for construction and operation, the geotechnical characteristics, potential environmental impact, regulatory permits related to use of land etc. are all considerations that factor into the decision. These variables can all be assessed through the use of GIS to narrow down viable options. Many of the features relevant to pipeline routing can be applied to siting facilities. For example, by taking surficial material, access, proximity to sensitive areas, and other criteria into account, facility locations can be determined through iterative scenarios and then confirmed by field visits. The goal of the desktop siting is to minimize the time and cost of field mobilization.

3.5.4.2 Civil works

Selecting a location for the pump or compressor station gives way to the creation of a plot plan where an arrangement of the equipment is laid out based on the process requirements. The layout will be developed in an iterative way as the detailed mechanical and civil engineering activities are completed. An overall layout provides an understanding of possible locations for foundations and piles for buildings and major equipment as well as access roads and other infrastructure.

When designing the foundations of major equipment such as pumps and compressors, it is important to take into consideration the possibility of vibration once the pumps and compressors start operating. Since this equipment provides the kinetic energy to either the oil or gas products, any vibration present would originate with this equipment. From the stand point of geotechnical analysis, understanding the type of soil and foundation requirements can provide a head start in dealing with the vibration issues. This information can be part of the GIS and models can be developed to simulate scenarios as well as determine road routing scenarios and cut and fill calculations.

3.5.4.3 Equipment – Electrical

This portion of the process starts with the utility transformer, and a determination of any necessary utility upgrades in order to provide power overall for the pump station, compressor station, meter station or tank farm. The road map for the electrical design is the single line diagram, which shows all the loads that can potentially draw power from the utility service. A GIS system may have detailed information on the type of power utility grids available for a given geographic area, whether a high voltage transmission line is in the vicinity of the proposed facility, or whether a mid or low voltage distribution system is nearby.

After defining the utility requirements, needs for a main bus, switch gear contactors and motors may be identified as equipment to be designed for the MCC (motor control center) building. Variable frequency drives allow pump and compressor motors to run at different speeds as the pipeline flow varies. Other ancillary equipment associated with lighting, heating, noise abatement, ventilation and air conditioning will also need to be specified.

The control system uses automation to manage how the facility operates locally or with some type of remote interface through a SCADA system (supervisory control and data acquisition). Locally the control system operates according to the control philosophy, pumps start and stop according to process variables such as suction and discharge pressures. Meters can send pulse information to a flow computer that corrects the volumes from process conditions to standard conditions. The control system usually interacts in a comprehensive way with the metering system to ensure an appropriate custody transfer of the commodity is taking place to and from the pipeline system. The control system also provides emergency shutdown sequences for different possible scenarios in order to protect the facilities, the line pipe, employees, the public and the environment. Loss of containment, in the case of hydrocarbons, is typically very harmful. The control system is intricately related to the instrumentation system as instruments are the eyes of the control system that allow it to make decisions based on pre-established logic.

The SCADA system provides the ability to execute commands remotely in order to ensure that a smooth operation takes place. The SCADA system involves several communication channels that can take the form of phone lines, radio communications, satellite communications and data concentrators that in combination allow remote interaction without having to have a man on site 24/7. The GIS can also help the documenting of SCADA links for the operation of the pipelines.

Detailed engineering for control systems, instrumentation and SCADA is evolving at a much faster pace when compared with the Civil, Electrical or Mechanical engineering disciplines. In the recent past, wireless communications inside facilities have reduced the costs of control cable installations. New technologies like Ethernet bridges, protocol converters and local networks are now available for local control systems to operate.

3.5.5 Facilities – Construction

A successful construction project starts with the procurement of major mechanical equipment and the appropriate set up of contracts with the different service providers. These contracts can be established with a reasonable level of commercial leverage and negotiating before being finalized when suppliers have the uncertainty of whether or not they will be awarded the contract.

Also involved is procurement of major equipment, construction materials such as pipe, cable, metal sheet, and consumables such as diesel. Sourcing construction equipment such as cranes, bulldozers, manlifts, grinders, diesel welders, flat beds, zoom booms, grading and compaction equipment and vacuum trucks is usually the responsibility of the construction contractor(s).

The logistics of when, and how to deliver pipeline equipment is driven by completion of the civil foundations and the readiness of platforms and concrete pads to receive them. The installation of mechanical equipment opens the way to installation of pipe racks and piping. Pipe racks and construction of MCC and Electrical buildings allows for installation of cable trays as well as the routing of cables from MCC buildings to motors and from control buildings to field devices and instrumentation.

At its current state, geomatics is limited in its support to facility construction, beyond assisting with legal and construction surveys. This is mainly because, unlike pipelines, facilities are not large geographic features and therefore can be designed and constructed without a great deal of spatially distributed data. Furthermore, design software like SmartPlant Enterprise are approaching facility data management in ways similar to GIS packages by storing data entities in a database rather than in drawings. Nonetheless, some areas are growing in application, such as high definition laser scanning (featured in Chapter 8) and construction monitoring which has been done from high resolution satellites (15).

3.6 What's Next?

Geomatics has introduced some real innovation to pipeline engineering and construction and has conversely greatly benefitted from the opportunities of working within a thriving industry. We are seeing geomatics play an increasingly important role in the design and construction phases of the pipeline life cycle.

Pipelines continue to grow while many more are aging, which require a constant accounting of their location, integrity, and potential impact on their surroundings. To help manage the volumes of data generated and maintained during this accounting, GIS provides many efficiencies. These systems keep data organized, secure, accessible, and effectively managed through continuous changes and updates. They also enable the automation and modeling of many engineering processes, thus providing efficiencies to pipeline projects and operations. With good foresight, geomatics should be part of a project from the beginning. In this way, data, processes, and data management can be carried through design, construction, and passed onto the client for operations and integrity.

Despite the gains made, there are still some areas where geomatics could play a larger role by leveraging its spatial indexing capabilities. Principal among these are:

- Pipeline project cost estimating – Given that routing and associated business rules can be housed and computed in a GIS, it is not a large step to add project cost values as data.
- Integration with facilities data and models – One of the biggest current gaps in facilities GIS is how geo-spatial data can be coordinated with specific plant design data.
- Data intensive calculations such as those using LiDAR data or 3D modeling or rendering are still taxing on current hardware. Video game technology is starting to play a larger role in dealing with this problem.
- Real-time data transfer to field devices and PDAs – Although this is currently possible and used by some companies, it requires an investment in infrastructure to support communications and the central database as well as new workflows for collecting, storing, and maintaining data.
- Central database implementation times – Setting up a central database on a project is becoming easier and easier. When it is as simple and common as pointing to a new project folder and as accessible as a website, then GIS will have become a common project practice.

References

(1) Kuhn, T.S., “The Structure of Scientific Revolutions”, 2nd. ed., Chicago: Univ. of Chicago Pr., p. 206, 1970

(2) Adam, S., “A Central Database Improves Pipeline Projects”, Proceedings of IPC: International Pipeline Conference, Calgary, Canada, 2006

(3) Guerrero, N., Brush, R., Bridges, G., Spickler, F., “Utilizing GIS to Manage your Assets”, GITA, GIS for Oil and Gas Conference 2008.

(4) Neurauter, S., Szeto, S., Tindall, M., Wong, Y., Wright, C., 3D “Visualization for Hydrocarbon Project Design”, Proceedings of IPC: International Pipeline Conference, Calgary, Canada, 2008

(5) Neurauter, S., Szeto, S., Tindall, M., Wong, Y., Wright, C., “Exploiting

the Visual Value of Digital Elevation Models", GeoTec Conference, Calgary, Alberta, Canada, 2007

(6) Adam, S., M. Farrell, "Earth Imagery Monitors Pipeline Integrity", Imaging Notes, 16(2): 18-19, 2001

(7) Riedmann, M., Thomas, A., Sims, R.A, Rogg C., Schleider, O., "Managing pipeline using satellite imagery", Pipeline and Gas Technology, 2009

(8) von Meyer, N., GIS and Land Records: "The ArcGIS Parcel Data Model", ESRI Press, 2004

(9) Hutson, A., "A Guide to Route Selection for Large Diameter Pipeline Projects", Proceedings of the Pipeline Division Specialty Conference 2006

(10) Cheng, M.Y., Chang, G., L., "Automating Utility Route Design and Planning Through GIS", Automation in Construction, Vol. 10, No. 4, pp. 507-516, Elsevier Science, 2001

(11) Rizkalla, M., "Pipeline Geo-Environmental Design and Geohazard Management," ASME, New York, 2008

(12) Adam, S., "Towards Permission To Construct Oilfield Facilities Without A Conventional Field Survey", Proceedings of IPC: International Pipeline Conference, Calgary, Canada, 2006

(13) An example of a jurisdictions pilot project is found at: LiDAR Pilot Projects for Well Sites (updated December 12, 2008):

http://www.srd.gov.ab.ca/lands/managingpublicland/landinformation/planinformation.aspx

(14) Refer to http://www.osha.gov/

(15) Kostianoy, A.G.; Ermakov, P.N.; Soloviev, D.M., "Complex Satellite Monitoring Of The Nord Stream Gas Pipeline Construction", Proceedings - US/EU-Baltic International Symposium, 2008 IEEE/OES, Page(s):1 – 5, 2008

4 Pipeline Integrity

4.1 Introduction

Pipeline integrity is a key element to the safe daily operation of a hazardous liquids or natural gas pipeline system. The amount of data and resulting information required by a pipeline operator to effectively manage the integrity aspects of their pipelines can be overwhelming. Pipeline integrity data encompasses basic pipeline properties, data related to integrity threats, data utilized in consequence modeling, the enormous amount of data obtained by inspections and surveys, and all attributes involved in a risk assessment. Successfully integrating all this pertinent information is key in managing pipeline integrity.

The pipeline industry is realizing that geomatics and geographic information systems (GIS) play an important role in adding value to integrity management. The gathering, storing, processing, and visualization of geographic information, or spatially referenced information, is conducted throughout every step of a pipeline integrity management plan. The analysis of accurate spatial data is crucial in all decisions made to maintain the safety of a pipeline system.

4.2 Pipeline Integrity Management

The primary objective of a pipeline operator is to manage the integrity of their pipeline system. Operators do this in order to provide safe, reliable and cost effective product delivery for their customers without adverse effects on the public, their customers, their employees and the environment. Pipeline integrity management is a process for assessing and mitigating pipeline risks in order to reduce the likelihood and consequences of incidents, which results in improved safety. This is accomplished by the development of an effective pipeline integrity management plan which provides a systematic, comprehensive, and integrated approach to managing the safety and integrity of pipeline systems.

A pipeline integrity management plan is a written document which is typically based on regulations outlined in federal or provincial/state legislation. These regulations require pipeline operators to identify areas of high risk that could be affected by an unintended release of the product transported by the pipeline. In order to define these high risk areas, pipeline integrity threats need to be identified and consequences related to failures due to these threats need to be analyzed. A pipeline risk assessment utilizes all the integrated information gathered about the pipeline and surrounding area to apply a risk score or rank to each dynamic segment. Pipeline operators are required to demonstrate that the risks associated with operating their pipelines have been assessed and are being managed.

Planning and executing an integrity program utilizes all this information in conjunction with appropriate field inspections and surveys. A pipeline integrity management plan must also include methods to measure the program's effectiveness in assessing and evaluating integrity. A GIS, utilizing effective spatial data management, plays a critical role in data collection, integration, analysis, and reporting to support a successful pipeline integrity management plan.

A summary of the general processes involved in a pipeline integrity management plan are outlined below:

- Data collection and integration
- Identification of pipeline integrity threats
- Definition of high consequence areas
- Perform a risk assessment
- Integrity program planning and execution (integrity assessments)
- Measure effectiveness, program improvement, documentation, feedback

These processes are outlined in the following work flow diagram (Figure 1).

The steps of an integrity management plan and how they relate to geomatics and GIS processes will be described in more detail throughout the following sections. Data sources and types will also be discussed in each section. To demonstrate how these processes and accompanying data work in real-world industry related situations, a number of pipeline integrity regulations and standards will be referenced. Many international regulations and standards are widely recognized and used as legislation for pipeline operators to produce an integrity management plan and to carry out a risk assessment. Examples of these regulations and standards are listed below.

- American Code of Federal Regulations, 49 CFR Part 192 Subpart O, Gas Transmission Pipeline Integrity Management
- American Code of Federal Regulations, 49 CFR Part 195.452, Pipeline Integrity Management in High Consequence Areas (Hazardous Liquid Pipeline Operators)
- American Society of Mechanical Engineers, ASME B31.8S - 2004, Managing System Integrity of Gas Pipelines
- American Petroleum Institute, API Standard 1160, Managing System Integrity for Hazardous Liquid Pipelines
- Canadian Standards Association, CSA Z662-07, Oil and Gas Pipeline Systems
- Australian Standard, AS 2885.3:2001, Pipelines - Gas and Liquid Petroleum - Operation and Maintenance

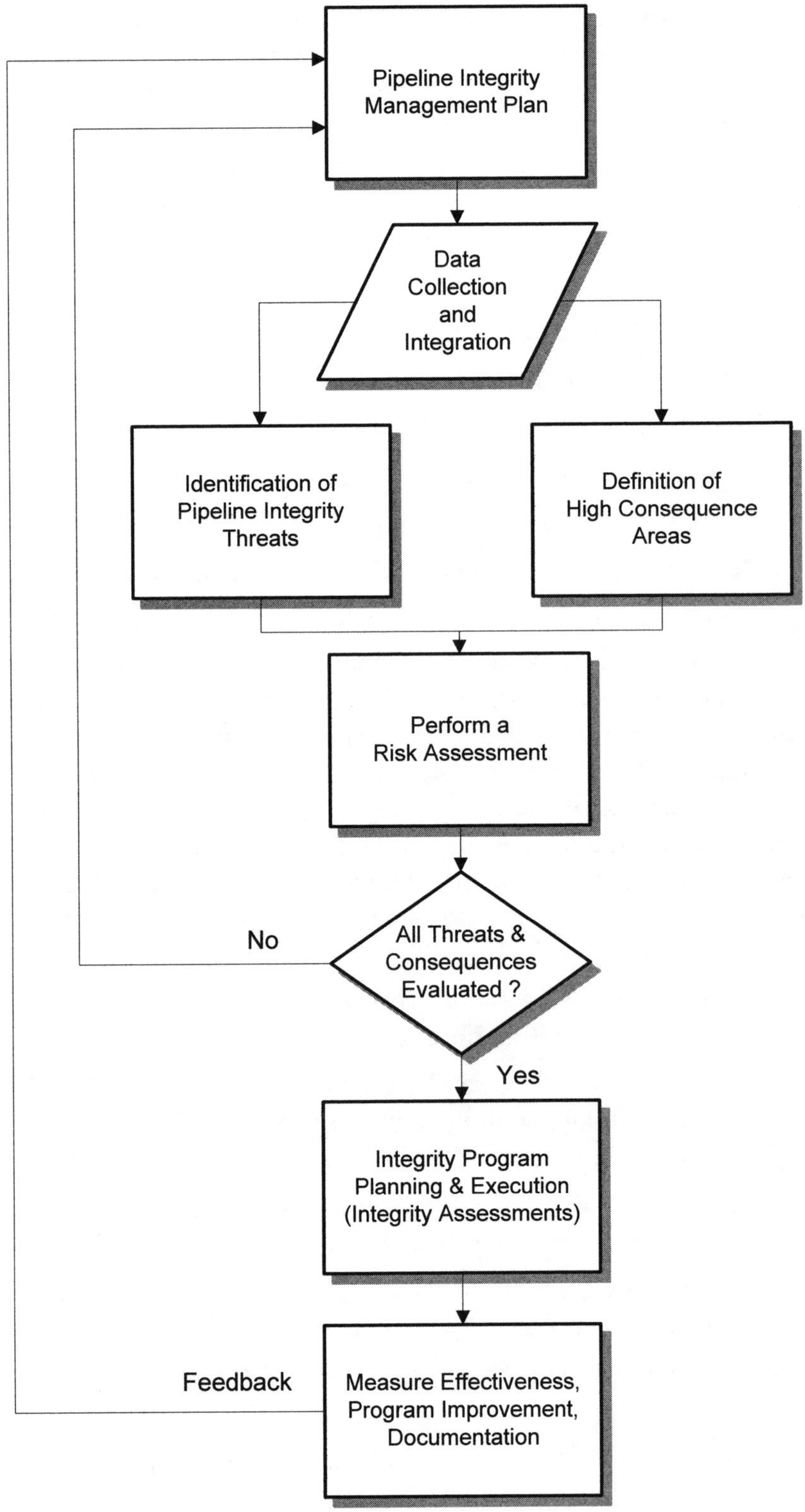

Figure 1 – Work Flow of a Pipeline Integrity Management Plan

- British Standard, BS PD 8010-1:2004, Code of Practice for Pipelines - Steel Pipelines on Land
- European Standard, BS EN 14161:2003, Petroleum and Natural Gas Industries - Pipeline Transportation Systems
- International Organization for Standardization, ISO 13623:2000, Petroleum and Natural Gas Industries - Pipeline Transportation Systems

There is no single country that outlines their current regulations adequately enough to deal with all possible aspects of every integrity management plan, but all the regulations stated above simply require conformance with clearly defined standards. Ultimately, it is the responsibility of the operating company to determine how they want to implement the requirements within the regulations and how to manage associated data records for future analysis or performance metrics purposes.

4.3 Data Collection and Integration

It is important to assess all data necessary to perform the processes of a pipeline integrity management plan. A data inventory is created and a strategy for collecting, maintaining, and visualizing data is required. This can be accomplished by implementing a GIS, databases, spreadsheets, and/or document management systems. A geomatics application or GIS is a natural choice for integrating pipeline integrity data, as there is a spatial component to most of the attributes involved. Once all necessary data has been collected, organized and stored, a GIS implementation facilitates spatial analysis to determine integrity threats, define areas of high consequence, and to create maintenance and remediation schedules. The information created by this spatial analysis is visualized using maps and documented with reports. The effective use of GIS applications streamlines the implementation of a pipeline integrity management plan and creates improved communication throughout all aspects of pipeline operations.

Data collection can be an overwhelming task for a pipeline operator, especially when they have acquired pipeline assets from other companies or if they have legacy data that may be in various formats and poorly organized. Typical legacy data may include paper documents, CAD alignment sheets, tabular spreadsheets and databases located across the organization, and various proprietary vendor data sources. Collecting and organizing this data can be a lengthy and involved process, but it is required as this builds the foundation of an effective GIS pipeline database. Well planned and effectively managed data conversion projects need to be considered by organizations wishing to move from legacy data management sources to an integrated GIS platform.

There are many factors to consider when integrating pipeline data into a data management system supported by a GIS. Since most data collected will have a spatial component, represented either by coordinates or a station location, the pipeline centerline plays a very important role. Accurately positioned centerlines with properly calibrated stationing are vital in correctly visualizing and analyzing all related pipeline attributes. Other things to consider during the data integration phase are:

- Data storage
- Data management systems
- Pipeline data models
- Integrating external datasets
- Consideration for subject matter expert data
- Understanding spatial data, coordinates and projections with their associated accuracies
- Choosing a GIS vendor, which may include data conversion and data maintenance software tools
- Utilizing field collected GPS data
- Aligning and integrating field surveys
- Performing spatial analysis
- Creating maps and reports
- Corporate and field data access

Industry specific vendor software is also available for pipeline integrity management. It is developed specifically for collecting and integrating pipeline integrity related data and converting it into information used for evaluating threats and consequences to calculate the risk value of a pipeline segment. Vendor software, in conjunction with an effective GIS and a comprehensive database, can streamline integrity management processes by integrating and characterizing many data sources into one common system.

4.4 Identification of Integrity Threats

After collecting and integrating all available data, the next step in developing a pipeline integrity management plan is identifying all potential integrity threats to the pipeline system. Incident history must be analyzed prior to determining these threats. The operator must be familiar with all incident types and causes on their pipeline system. Using documented historical information in conjunction with available third party research on pipeline threats will effectively identify all integrity threats applicable to an operator's pipeline system.

An example of independent research leading to a standard on pipeline integrity treats is outlined in ASME B31.8S, Paragraph 2.2 (1). Here gas pipeline incident data has been analyzed and classified by the Pipeline Research Committee International (PRCI) into 22 root causes. Each of the 22 causes represents a threat to pipeline integrity that should be managed. These threats are grouped into four types and summarized in Table 1:

- Time dependent threats, such as internal corrosion, external corrosion, and stress corrosion cracking
- Static or resident threats, such as fabrication or construction defects
- Time independent threats, such as third party damage, incorrect operations, and outside force damage

Table 1 - ASME B31.8S Threat Classification

Defect Type	Defect Category	Threat
Time Dependent	External Corrosion	External Corrosion
	Internal Corrosion	Internal Corrosion
	Stress Corrosion Cracking	Stress Corrosion Cracking
Stable	Manufacturing Related	Defective Pipe Seam
		Defective Pipe
	Welding / Fabrication Related	Defective Pipe Girth Weld
		Defective Fabrication Weld
		Wrinkle Bend or Buckle
		Stripped Threads / Coupling Failure
	Equipment Related	Gasket O-Ring Failure
		Control / Relief Equipment Malfunction
		Seal / Pump Packing Failure
		Miscellaneous
Time Independent	Third Party / Mechanical Damage	Damage Inflicted by First, Second, or Third Parties (instantaneous failure)
		Previously Damaged Pipe (delayed)
		Vandalism
	Incorrect Operations	Incorrect Operational Procedure
	Weather Related and Outside Force	Cold Weather
		Lightning
		Heavy Rains or Floods
		Earth Movements
Unknown	Unknown	Unknown

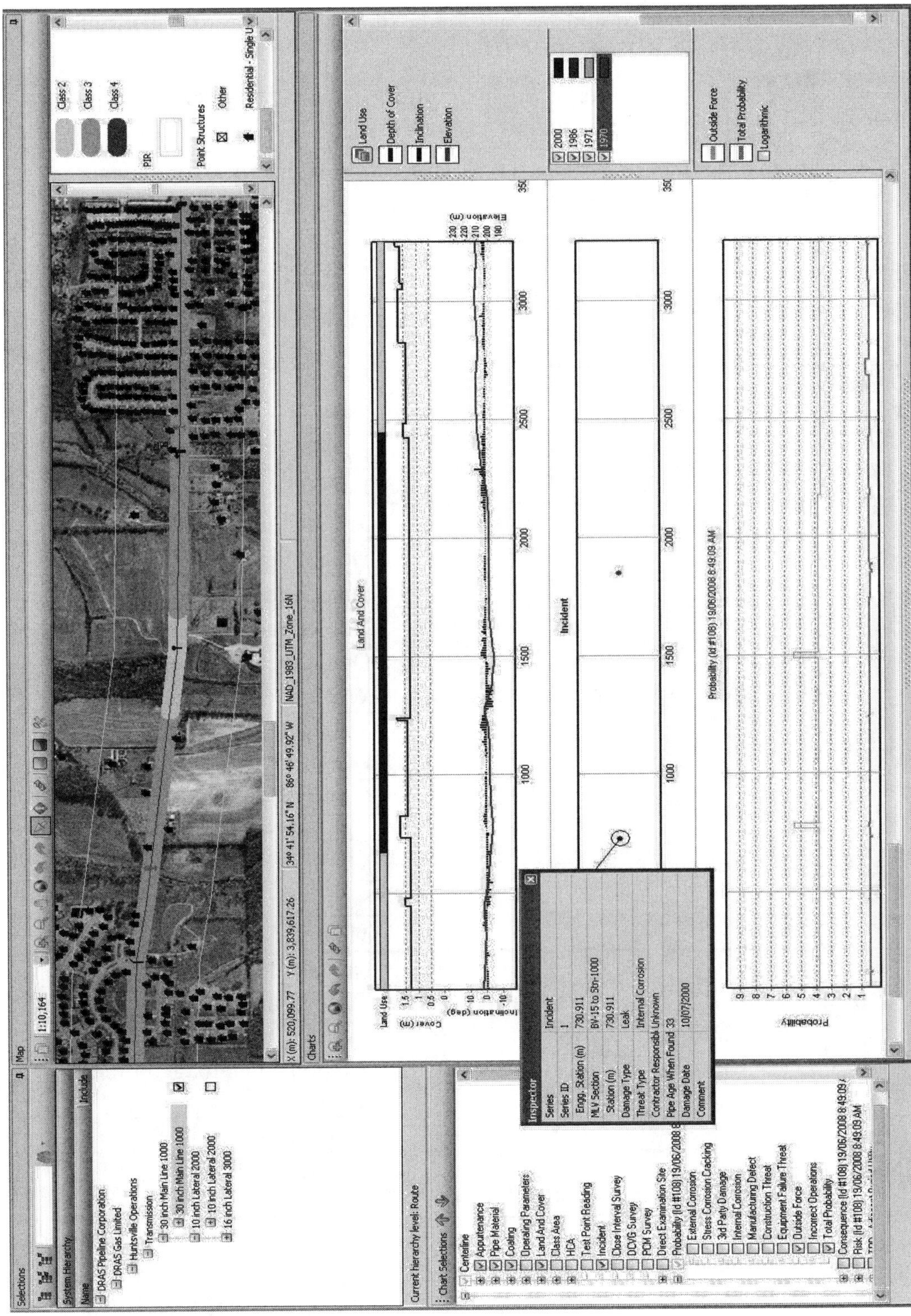

Figure 2 – Using a GIS Application to Manage Third Party and Outside Force Threats

- Unknown threats, where no root cause or causes have been identified

Examples of datasets that can accomplish the goals of a public awareness or pipeline surveillance program include the following:

- Land based data (land tiles, ownership, address points)
- Occupancy data (occupied structure points and polygons representing areas of public gathering)
- Municipal, county or state boundaries
- Emergency responders (locations, response times, zones of coverage)
- Urban development or land use planning information (rezoning applications, subdivision applications, restrictive covenants)
- Roads, highways, navigable waterways
- Pipeline incidents, risk assessment and failure history data
- Pipeline crossing agreements (road use, crossing agreements, leases)
- Pipeline surveillance program

The application of a GIS to both house multiple data sets and assist in ongoing analysis can be instrumental in the effectiveness of a public awareness and associated damage prevention programs, such as pipeline surveillance (Figure 2).

4.5 Definition of High Consequence Areas

Pipeline operators must know the characteristics and locations of areas of high consequence related to their pipeline system to make decisions required by their integrity management plan. The techniques used in determining pipeline segments influenced by areas of high consequence vary depending on whether the product type is gas or liquid. Even though these techniques vary, spatial analysis, accurate data, and an effective GIS play a crucial role in defining high consequence area (HCA) segments for pipeline operators.

Consequence modeling varies depending on the product type of the pipeline. Gas and liquid are the two general product types that will be considered for determining HCA segments on transmission pipelines. Gas transmission operators are required to analyze how a thermal impact circle caused by a potential pipeline failure relates to population density and locations of buildings and areas intended for human occupancy. In comparison, liquid transmission operators must model releases from their pipeline system into predefined areas of high consequence, such as ecologically sensitive areas, water bodies, populated areas, and sources of drinking water. The American Federal Codes 49 CFR Part 192.903 (gas) and 49 CFR Part 195.450 (liquid) are two examples of how areas of high consequence are defined and modeled in the pipeline industry. These two standards will be referenced throughout this section. Other approaches such as modeled techniques to determine hazard radii due to a product release and subsequent ignition would similarly be used to analyze the intersect between a failure impact zone and populated areas or other environmental receptors.

4.5.1 Gas Consequence Modeling

The complete definition of the gas HCA rule with calculations and parameters can be found in the published document, 49 CFR Part 192.903 (3). Gas transmission pipeline operators have the option to choose two methods to identify HCA segments, Method 1 and Method 2.

Method 1 is focused on defining Class locations along the pipeline. Class definitions are essentially related to population density and are determined by counting buildings and areas intended for human occupancy. These structures are tallied by sliding a class location unit (CLU) buffered 220 yards (200 meters) on either side of the centerline of any continuous 1-mile (1.6 kilometers) length of a pipeline (Figure 3). The complete definition of Classes 1 through 4 can be referenced in 49 CFR Part 192.5 (3). An HCA segment defined by Method 1 consists of a Class 3 or Class 4 location. In addition, an HCA includes any Class 1 or Class 2 locations where the potential impact radius is greater than 220 yards (200 meters), and the area within a potential impact circle contains 20 or more buildings intended for human occupancy. Any area in a Class 1 or Class 2 location where the potential impact circle contains an identified site also defines an HCA segment.

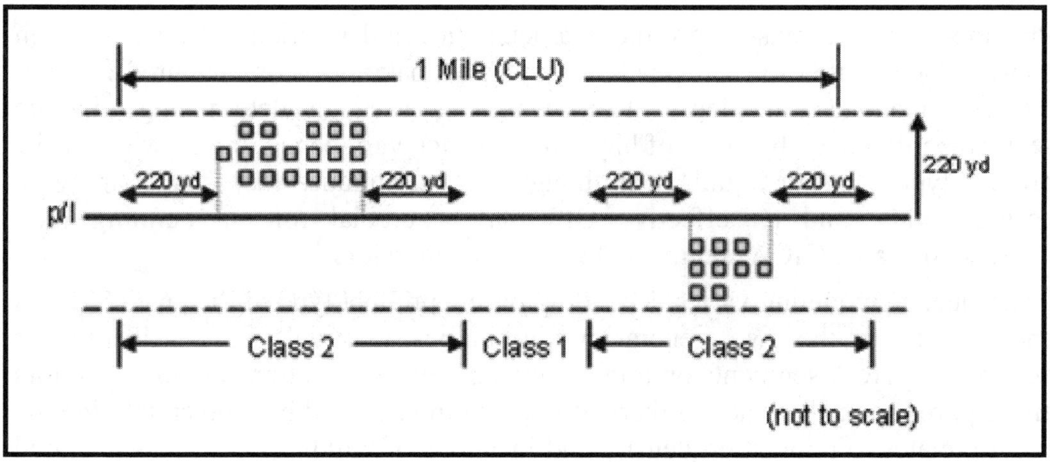

Figure 3 – Class Determination used for Method 1 Gas HCA Modeling

Identified sites are defined as an outside area that is occupied by 20 or more persons on at least 50 days in any 12 month period, a building that is occupied by 20 or more persons on at least 5 days a week for 10 weeks in any 12 month period, or a facility occupied by persons who are confined due to impaired mobility or would be difficult to evacuate. Some examples of identified sites are playgrounds, stadiums, churches, office buildings, hospitals, prisons, and retirement facilities.

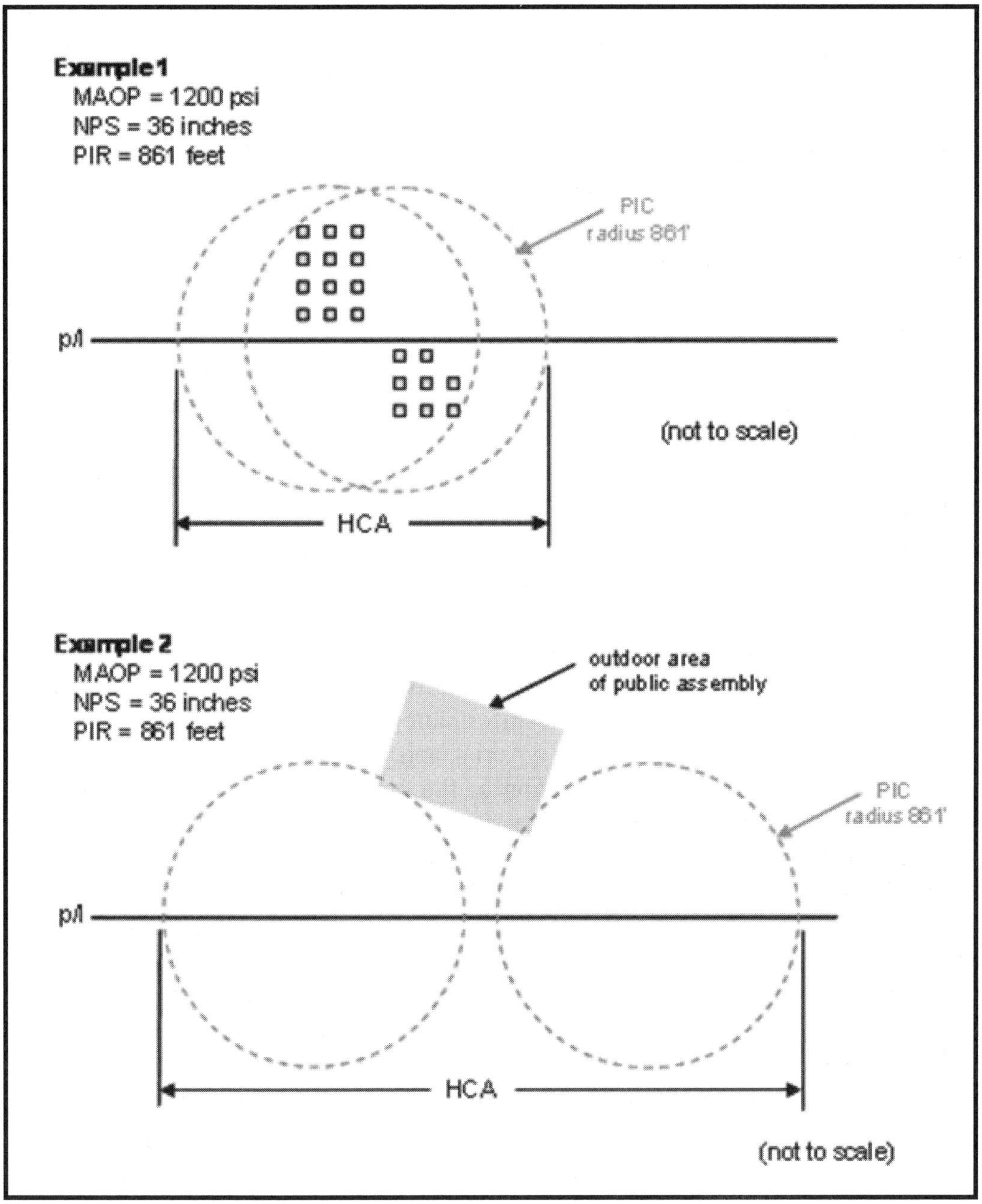

Figure 4 – Method 2 HCA Determination

The potential impact radius (PIR) means the radius of a circle within which the potential failure of a pipeline could have significant impact on people or property. PIR is determined by the formula r = 0.69* (square root of (p*d2)), where 'r' is the radius of a circular area in feet surrounding the point of failure, 'p' is the maximum allowable operating pressure (MAOP) in the pipeline segment in pounds per square inch and 'd' is the nominal diameter of the pipeline in inches. A potential impact circle (PIC) is defined by a radius equal to the PIR.

Method 2 in determining gas transmission HCA segments does not involve Class locations. It generally states that an HCA along the pipeline is defined by a potential impact circle containing 20 or more buildings intended for human occupancy or includes an identified site. Figure 4 shows two examples of the spatial analysis involved with determining HCA segments by Method 2.

The methodology outlined in 49 CFR Part 192.903 (3) to determine pipeline segments affected by areas of high consequence is well defined by a logical step-by-step approach. There is a significant effort involved in performing these calculations manually, especially when pipeline operators are dealing with large cross-country transmission systems. Since all of the related data has a spatial component, the effective use of GIS technology leads to an automated, repeatable, and defendable approach to calculate gas transmission HCA segments in an efficient manner.

The quality of the spatial data used is crucial in determining accurate gas HCA segments. The most important factor in this analysis is the pipeline centerline. The Class locations, structures, and identified sites (structure points or outdoor areas) are located and assessed based on the buffered centerline, therefore it is important to use all available source information to ensure the highest level of accuracy of the pipeline centerline. Aerial and satellite photography is a very useful data source to verify the location of the pipeline centerline. The pipeline right-of-way is typically evident in the images and in some cases above ground appurtenances are visible as well. Survey location points taken along the pipeline from sub-meter global positioning system (GPS) equipment can be used to verify or modify the centerline. Most operators can not justify initiating a program to obtain GPS coordinates for their entire pipeline system, so it is important to utilize GPS technology whenever a field crew is conducting work on the pipeline. This includes close interval CP surveys, depth of cover surveys, right-of-way patrols, and pipe inspections. An inline inspection (ILI) utilizing an inertial mapping unit (IMU) is another way to obtain accurate GPS coordinates for piggable pipelines.

Accurately digitized buildings and areas along the pipeline is also critical to properly define Class locations and determine the HCA segments utilizing the potential impact circle. A point or polygon feature can be used when incorporating these buildings and areas into a GIS. Polygon features are obviously the most accurate representation of all sites used in the HCA analysis, but in some cases, due to time and resource constraints, digitization of single family dwellings are digitized using point features (either the centroid of the building or the point of the structure that is closest to the pipeline).

The calculations and spatial analysis involved in defining Class locations and gas HCA segments is relatively simple, but to perform an efficient systematic analysis throughout a large transmission pipeline system may require some customizations to off-the-shelf GIS applications. There are a number of third-party vendors offering pipeline operators custom software applications to leverage an existing GIS and pipeline database to make the determination of Class and HCA segments automatic.

4.5.2 Liquid Consequence Modeling

The standards 49 CFR Part 195.452 (4) and API 1160 (5) give guidance on how to define pipeline segments that could affect areas of high consequence for liquid systems. As stated in Chapter 6 of API 1160, a liquid high consequence area (HCA) is a location where a pipeline spill might have significant adverse impacts to the surrounding population, environment, or navigation of commercial waterways. This includes spills from pipeline segments outside a direct intersect HCA that may be transported and consequently affect an area of high consequence a distance away from the pipeline.

The inspection protocols outlined in these standards require the operator to consider the release locations and volumes of a potential pipeline failure, the overland spread of the spill, the water transport of hazardous liquids, and the potential effects of the release. API 1160 Chapter 6.3 states that pipeline operators should consider the following when determining a potential impact zone:

- The health and safety consequences of a release, including the possible need for evacuation.
- The nature and characteristics of the product or products transported (refined products, crude oil, highly volatile liquids, etc.).
- The operating conditions of the pipeline (pressure, temperature, flow rate, etc.).
- The topography of the land associated with the HCA and the pipeline segment.
- The hydraulic gradient of the pipeline.
- The diameter of the pipeline, the potential release volume, and the distance between isolation points.
- The type and characteristics of the HCA crossed or in proximity to the segment.
- Potential physical pathways between the pipeline and the HCA.
- Potential natural forces inherent to the area (flood zones, earthquake zones, subsidence areas, etc.).

- Response capability (time to detect, confirm and locate a release; time to respond; nature of response; etc.).

Developing an understanding of the potential impact zone ensures that operators include the appropriate HCA information for use in their risk assessment.

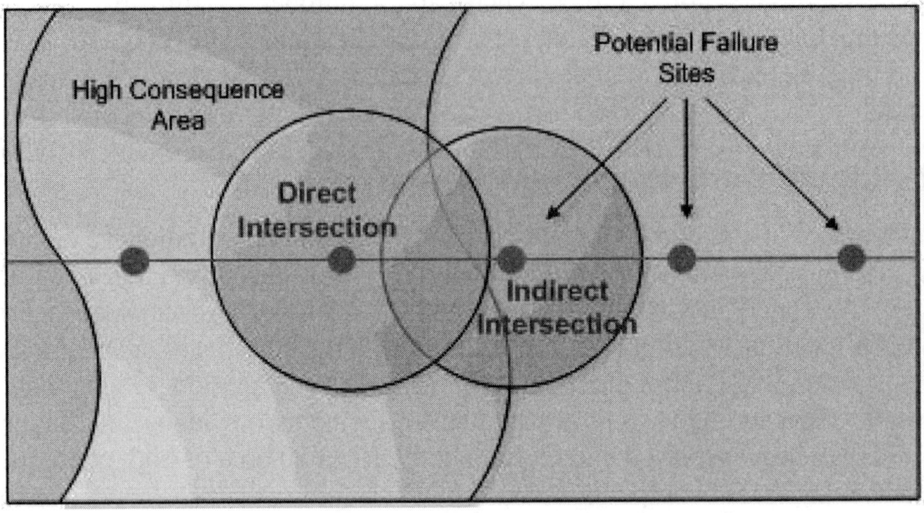

Figure 5 – Direct and Indirect HCA Intersections

The process for identifying segments that impact HCAs requires that pipeline segments and facilities physically located within the HCA are identified and defined by specific locations where the pipeline intersects the HCA boundary. These segments are commonly referred to as "direct intersect" segments. Segments physically located within a HCA may only be excluded if an operator provides a technical justification (risk assessment) that the segment could not affect the HCA. Pipeline segments located outside of the HCA which could potentially migrate to and indirectly affect an HCA are also identified and defined by specific begin and end points. These segments are commonly referred to as "indirect" or "could-affect" segments. Figure 5 demonstrates how direct and indirect intersections are determined.

As a cross-country pipeline travels through a number of geographically different areas and environments of different sensitivities, it becomes significantly more difficult for engineers to accurately justify the relative severity of any of these failure impacts using only experience and hand calculations. In light of this, hazardous liquid consequences from a pipeline failure are best quantified and mitigated through prioritized planning with the help of a computer model utilizing spatial analysis and a GIS (6).

Identifying direct segments is a simple process which may be accomplished using spatial tools. However, there is no well-defined procedure for identifying could-affect segments. Typically most approaches will fall into one of two categories: defining a spill buffer zone or performing a spill transport analysis. Both approaches would be expected to consider the API 1160 criteria mentioned previously. Depending on the nature of the product being transported, the process for identifying could-affect segments that impact HCAs will consider the transport of the spill by air, land or water. For either approach, spatial data is an important element of the analysis which should consider the surrounding terrain and proximity of the pipeline to water crossings.

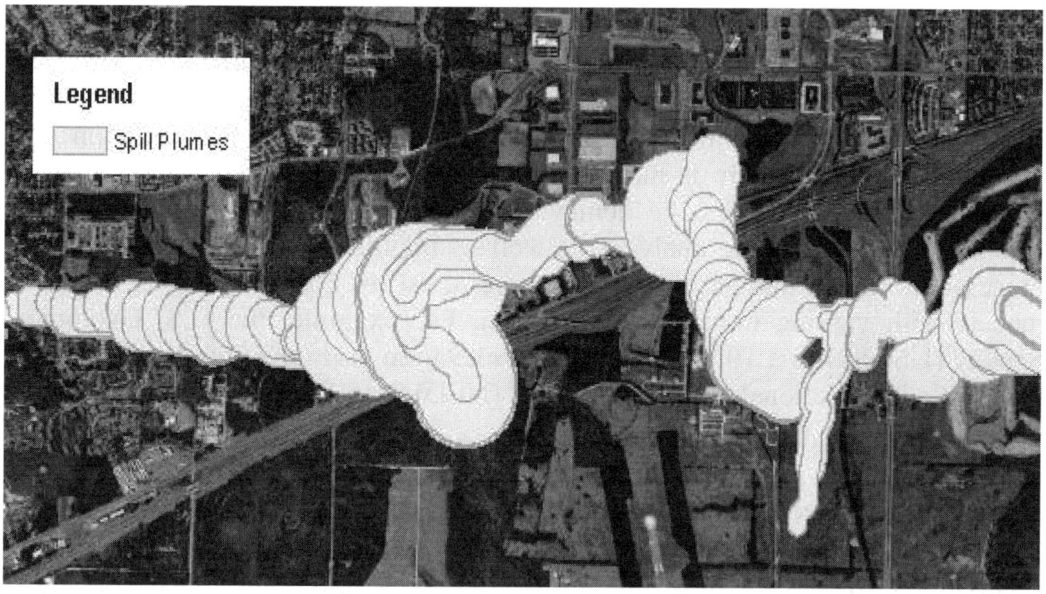

Figure 6 – Spill Plumes Created by an Overland Spill Analysis

In the first approach, the defined spill buffer distance is based on an estimated overland spread distance for a particular area. The second approach of overland spread analysis has an advantage in that it can estimate the distance and direction of the spill and determine whether the extent of a spill may impact an HCA for each particular release location. The overland spread analysis should consider the surrounding elevation changes and local factors, such as ditches or drain tiles, in determining the path of a spill. Figure 6 shows an example of spill plumes created by an overland spread analysis.

The volume of commodity that could be released given a pipeline failure is important in determining the potential effects and consequences. The larger the volume released, the greater the area that is impacted by the failure. The spill volume given a potential failure along the pipeline may be estimated from a drain down analysis utilizing a pipeline elevation profile. Local elevation changes are an important factor to take into consideration for estimating the volume that drains from the pipeline. An overland spread analysis, leveraging spill release volumes and elevation data, can then be used to determine the migration paths. In determining pipeline segments affecting areas of high consequence, it is obvious from these methods that spatial data plays a very critical role.

Similar to gas consequence modeling, an important factor in this analysis is the pipeline centerline. The spill buffer or overland spread is based on release locations along the pipeline centerline. An inaccurate pipeline centerline will affect the proximity of the spill to high consequence areas, as well as affect the results of a detailed overland spread analysis which relies on the local elevation changes surrounding the release location.

The elevation data is also a significant factor in a detailed analysis approach, which requires high resolution data in order to capture the changes in the landscape which would affect the migration of a spill. As the elevation grid resolution increases, there is more data to be processed in a computer analysis. Therefore, a balance must be achieved to obtain data with sufficiently high resolution to perform the analysis, as well as maintain a fast enough computer processing time. A number of public sources exist for obtaining digital elevation models with varying degrees of resolution. For example, the U.S. Geological Society (USGS) has a 10 meter resolution elevation grid covering the continental USA called the National Elevation Dataset (NED) (http://ned.usgs.gov/).

The locations classified as high consequence areas for hazardous liquids in the United States are provided to pipeline operators through the National Pipeline Mapping System (NPMS). The NPMS is a GIS created by the U.S. Department of Transportation (DOT), Pipeline and Hazardous Materials Safety Administration (PHMSA), Office of Pipeline Safety (OPS) in cooperation with other federal and state governmental agencies and the pipeline industry, as a tool to support the pipeline industry with compliance to federal regulations (http://www.npms.phmsa.dot.gov/). It is the responsibility of the operator to review other information sources, such as aerial surveys or census data, on a regular basis to supplement the NPMS data.

Additional datasets may be employed depending on the methodology for determining overland spread. Detailed and accurate hydrography data may be used to supplement the NPMS waterway data. A detailed hydrographic network can be used for performing a water transport analysis if a spill reaches a water resource. Soil and land cover information can also assist in determining the extent that a spill will spread over land.

The revisions of HCA boundaries and their associated pipeline segments represent an ongoing task. A repeatable procedure is required to review the system on a regular basis for any potential changes. The use of GIS technology is enabling operators to develop an effective procedure for consequence modeling on their pipeline system and achieve compliance with federal regulations.

4.6 Risk Assessment

When data collection and integration is complete, pipeline integrity threats determined, and areas of high consequence defined, a risk assessment can be initiated. A risk assessment supports pipeline integrity management initiatives by identifying pipeline segments within a system that are operating at the highest risk, and which are therefore, most in need of inspection and maintenance. The management of pipeline integrity on the basis of risk minimizes operational risks while optimizing the expenditure of resources in controlling risk.

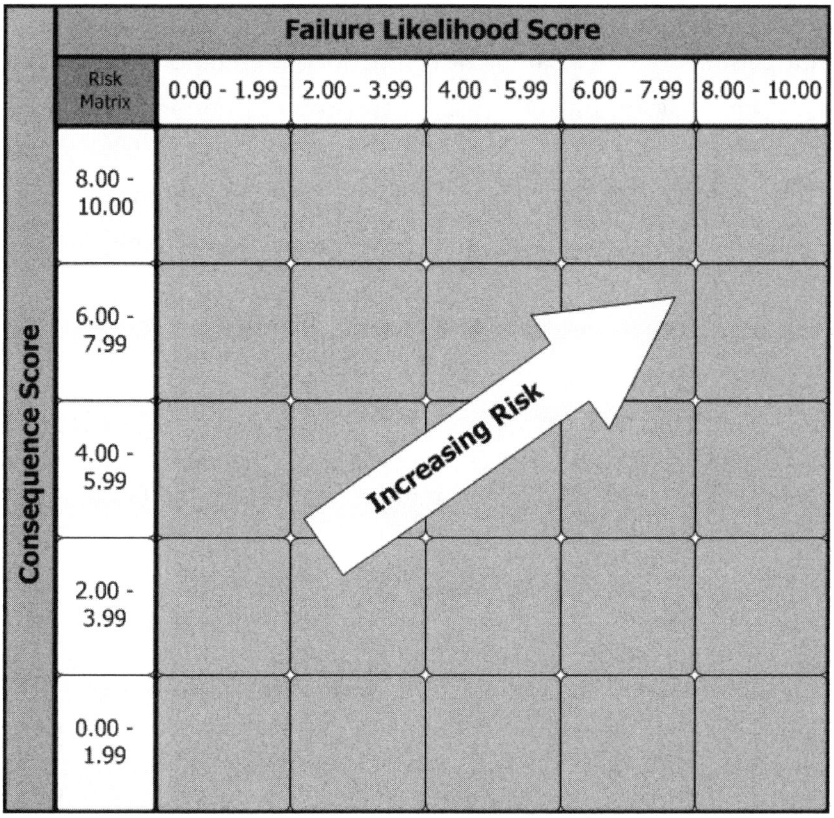

Figure 7 - Risk Matrix Format

Risk can be expressed as the joint consideration of likelihood of failure of a pipeline segment and the consequences of such failure. It can therefore be expressed in terms of the product of failure likelihood (or probability) and consequences, or it can be represented on a risk matrix, as illustrated in Figure 7.Utilizing the results of the failure likelihood and consequence scoring methodologies, a "5 x 5" risk matrix can be employed as the means for communicating the overall relationship between the failure likelihood score and the consequence score for each pipeline segment. Pipeline segments are grouped within the predetermined ranges of consequence scores and failure likelihood scores. These segments are then displayed in the appropriate matrix cell. The risk matrix provides the operator a simple visual representation of the distribution of risk scores for their pipeline system. An example of a risk matrix displaying a typical segment distribution is shown in Figure 8.

Risk assessments are used within the framework of the integrity management plan to:

- Identify the highest risk pipeline segments;
- Prioritize pipeline segments on the basis of risk;
- Identify primary risk-drivers so that appropriate assessment and risk mitigation strategies can be developed for each pipeline segment;
- Develop risk-optimized risk mitigation activities through "what-if" analyses;
- Determine the effectiveness of previous risk mitigation activities;
- Evaluate the impact of modified inspection intervals on pipeline integrity;
- Assess the use of, or needs for alternative inspection methodologies; and,
- Facilitate effective resource allocation

The risk assessment methodology common to all pipeline integrity regulations and standards is designed to provide a comprehensive, systematic and integrated risk management tool, addressing all potential threats, and facilitating rational decisions on a consistent basis.

The risk assessment methodology incorporates dynamic segmentation, as defined in ASME B31.8S, Section 5.7 k (1). An effective risk assessment process shall incorporate sufficient resolution of pipeline segment size to analyze data as it exists along the pipeline. Such analysis will facilitate location of local high-risk areas that may need immediate attention. For risk assessment purposes, segment lengths can range from units of feet (meters) to miles (kilometers) depending on the pipeline attributes, its environment, and other data. Manual creation of dynamic segments on a large transmission pipeline system is practically impossible due to the number of attributes considered and the total length of the pipelines assessed.

A fully loaded and verified database utilizing a standard pipeline data model should be a prerequisite to determining the dynamic segments of a pipeline system. Certain pipeline attributes are key to risk assessment calculations and are required to be linear continuous features. A linear continuous feature is defined by the following statements:

- A record for the data attribute must be provided for every location on the pipeline.

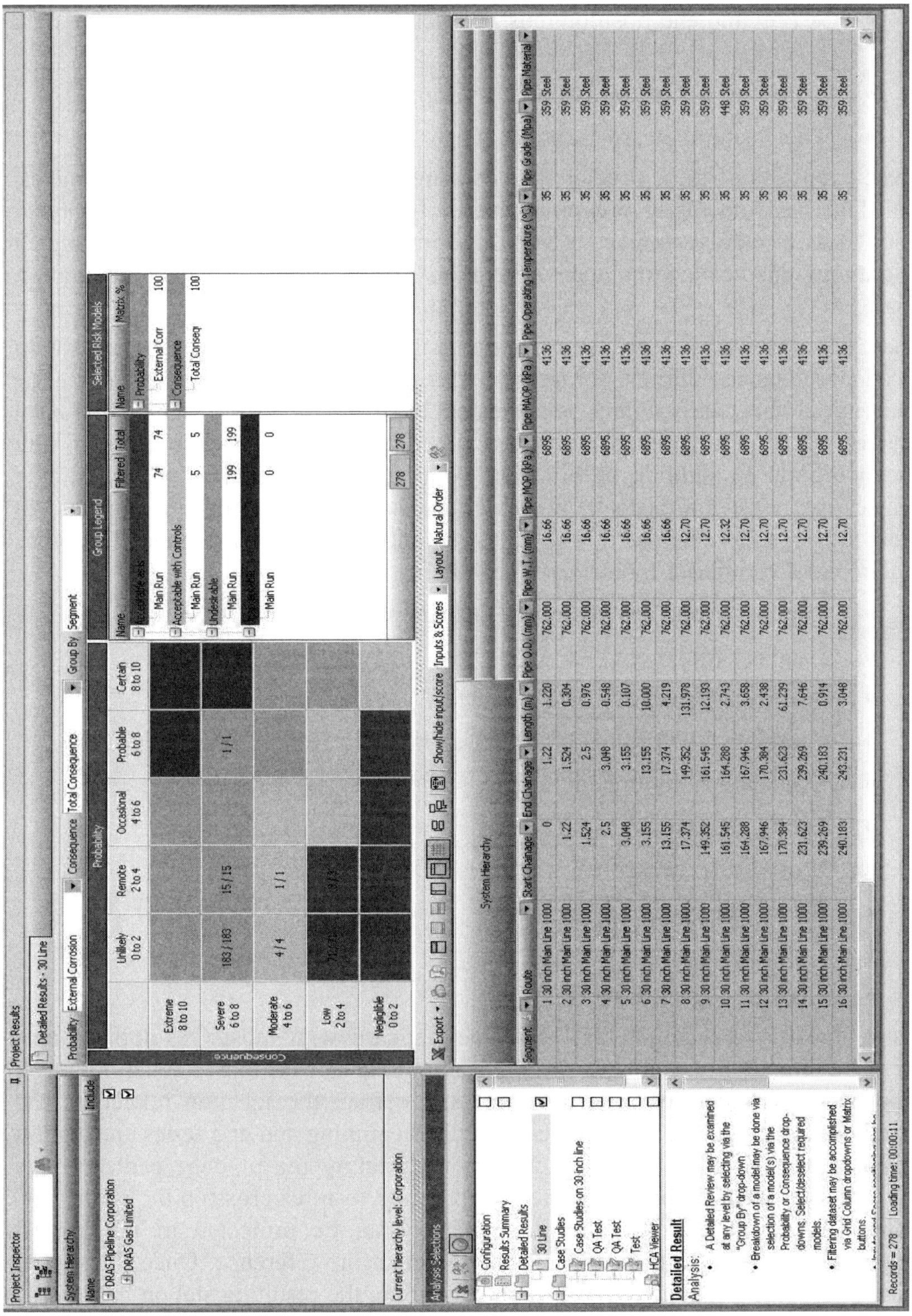

Figure 8 – Risk Matrix Showing Typical Segment Distribution

- A start and end position must be used to specify the location of each data record.
- Data attribute records cannot overlap.
- There cannot be any locations where no data is provided.

Some examples of these crucial attributes are date of pipe installation, diameter, wall thickness, operating pressure, grade (pipe specification), and coating type. These data elements would rank very high in importance and should be assigned priority in data collection and verification efforts during the data integration phase.

These key risk attributes are typically stored in a pipeline database as records with location references defined by begin and end series and station numbers. The majority of this tabular data was created by extracting from non-spatial data sources, such as alignment sheets, as-built documents, inspection records, and survey reports. In addition to this tabular data, there are a number of spatial data sources that are required to be processed and loaded into a pipeline database in order for risk to be calculated. The following list outlines several spatial data types required to determine risk on a pipeline system:

- Population and environmental receptors (points, polygons)
- Land use (polygons)
- Soil type (polygons)
- Crossings, such as roads, railways, utilities, rivers, creeks (polylines)
- Digital elevation model (grid)
- Fault lines (polylines)
- Frost depth (polygons)
- Seismic areas (polygons)
- Unstable slopes (polygons)
- One-call locations (points)

Using linear referencing tools that come standard with most GIS applications, there are simple procedures to overlay spatial data sets on top of a station aware pipeline centerline (measured polyline) and extract the location related to the pipeline stationing. For polygon features, the beginning and end series and station numbers can be extracted for each area intersecting the pipeline centerline. A station based intersection is obtained for polyline features crossing the centerline. Point features in proximity to the pipeline can be projected to the closest perpendicular distance to acquire a station centerline reference. Once the spatial data is processed against the measured centerline, the resulting station based data can be easily loaded into the tabular structure of a standard pipeline database.

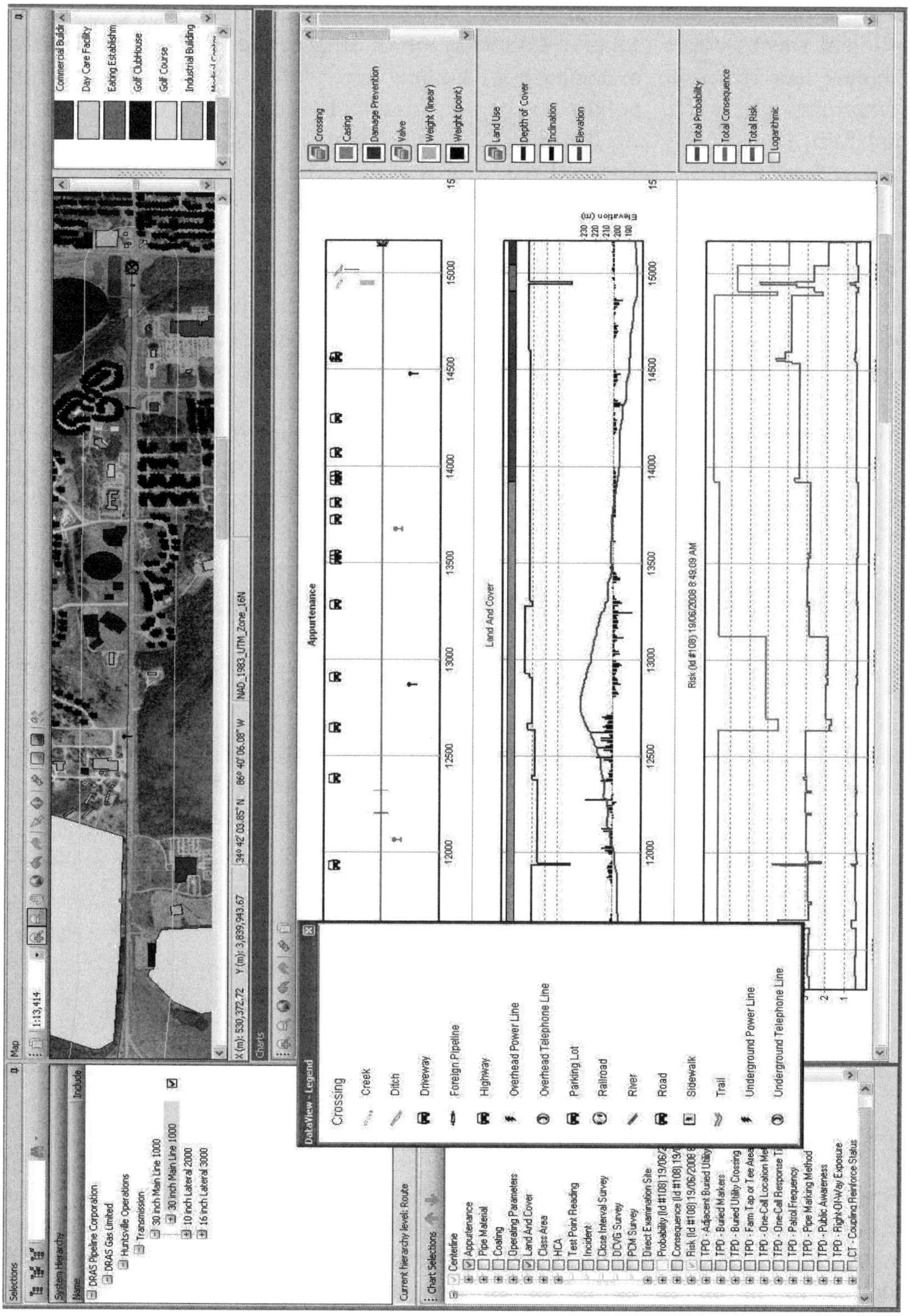

Figure 9 – Visualization of Dynamic Segments

When a pipeline database is fully populated, a computer application is the most efficient way to create dynamic segmentation of all pipeline attributes. Utilizing tabular data stored in a database in conjunction with station aware pipeline centerlines (measured polylines) creates a simple and effective process to transform linearly referenced data into spatial features that can be displayed and analyzed on a map. Calculated risk scores and supporting pipeline data can be symbolized in a GIS using various line types and colors. This can be done for individual threats and consequences, or total risk scores can simply be visualized (Figure 9).

A number of risk analysis software applications are available to the pipeline operator. These offer visualization of pipe properties and calculated risk scores in an interface showing data bands (similar to alignment sheets) synchronized with a map view. Being able to visually see the dynamic segmentation of the risk scores in relation to spatial attributes of the pipeline and surrounding area is extremely beneficial to the integrity engineer. These custom software applications facilitate in identifying the highest risk pipeline segments, prioritizing these segments, and determining the appropriate assessment and risk mitigation strategies.

4.7 Integrity Program Planning and Execution

A risk assessment provides information to evaluate both the potential consequence of different incidents and the likelihood that these events may occur. Having this information supports the integrity management process by allowing rational and consistent decisions. Risk results are used to identify locations for integrity assessments and resulting mitigation. GIS technology plays a key role in planning and executing an integrity program to reduce risk in a pipeline system.

A pipeline integrity program can consist of the following three types of assessment methods.

- Inline Inspection (ILI) Tools
- Direct Assessment
- Hydrostatic Testing

ILI tools and direct assessment will be addressed in this section due to the spatial data involved in these methods.

4.7.1 Inline Inspection (ILI) Tools

Pipeline integrity can be determined though an inline inspection (ILI) tool, often called a "smart pig", which is run internally through the pipeline to detect loss of metal and in some cases deformations in the pipeline. There are three types of ILI tools that help pipeline operators make integrity decisions about their pipelines: metal loss tools, crack detection tools, and geometry tools.

Another ILI tool that utilizes geomatics technology is a mapping tool. It is typically run in conjunction with other tools and uses an inertial mapping unit (IMU) to provide global positioning system (GPS) coordinates for the pipeline centerline, appurtenances, and defects. The GPS coordinates obtained from the mapping tool simplifies the process of calibrating the ILI data onto the proper location of the pipeline centerline. These coordinates can be loaded into a GIS application, overlaid onto a station aware centerline, and processed to extract the corresponding series and station number reference for each ILI feature.

If the ILI run did not implement a mapping tool, then aligning the resulting data with the pipeline stationing is slightly more complicated and involves spatial processing. The ILI data can be fit onto a centerline utilizing a "rubber-banding" technique using features that are common to the ILI run and the pipeline. Examples of these control point features are above ground markers, valves, tees, taps, and casings. To get an accurate fit of the ILI data, these features should be located with sub-meter accuracy GPS equipment.

4.7.2 Direct Assessment

Direct assessment, as stated in ASME B31.8S Section 6.4 (1), is an integrity assessment method utilizing a structured process through which the operator is able to integrate knowledge of the physical characteristics and operating history of a pipeline system or segment with the results of inspection, examination and evaluation in order to determine the integrity. Direct assessment methods include excavations where the pipeline is uncovered and the external coating and the steel pipe is inspected, as well as various types of cathodic protection surveys, such as close interval surveys.

Integrating a GIS with the planning phase of a direct assessment excavation produces an efficient method of accurately locating the dig site in the field. Overlaying the centerline location of the target anomaly with base layers such as aerial imagery, buildings, roads, and water bodies creates a detailed map to aid in navigation. The longitude and latitude coordinates of the target anomaly, used in conjunction with this map, gives a GPS enabled field crew the precise location to excavate. Obtaining GPS coordinates of the start and end of the excavated pipe and any girth welds exposed will give excellent positional reference to all pipe defects measured relative to a girth weld. This is an important process, as the excavation data will be loaded into the pipeline database based on stationing.

A close interval survey (CIS) is a technique used as part of an external corrosion direct assessment program that utilizes spatial data. This survey assesses the cathodic protection (CP) effectiveness over a length of a pipeline. GPS synchronized data loggers are used to record on and off potential readings at intervals of 1 to 3 meters on a pipeline cathodically protected by an impressed current CP system. If all CIS readings have associated GPS coordinates, then this makes the process of fitting this survey on a pipeline centerline very easy. The reading coordinates can be loaded into a GIS application, overlaid onto a station aware centerline, and processed to extract the corresponding series and station number. If no or sporadic coordinates are related to the CIS readings, then a "rubber-banding" technique similar to fitting ILI data between physical match points must be utilized. Features recorded in the close interval survey that are common to GIS base layers or that can be identified from imagery are used as control points in the aligning process. Typical control points during a close interval survey are edges of roads, edges of water bodies, driveways, and test leads.

4.8 Future Trends

Even though the pipeline industry has been in existence for a very long time, the industry has been slow to fully utilize the technology of geomatics and GIS for pipeline integrity. Many pipeline companies still use hard copy maps for managing and monitoring their pipelines or have only recently begun initiatives to develop a GIS and database for pipeline integrity and risk assessments. In order to satisfy regulatory compliance requirements, many pipeline operators are being forced to review their tabular and spatial data for completeness and accuracy.

Integrity management can be a challenging task as multiple stakeholders work to share information efficiently. In most cases, without a structured approach, data sets are incomplete, lack validation and are unable to be integrated. This ultimately results in programs that can't achieve their objectives. A programmatic approach to collecting, validating, integrating and analyzing integrity management information is critical to enable focused programs that are efficient and deliver value to the organization.

Standardized data collection for a pipeline integrity management program is greatly increased with the use of mobile mapping tools. A GIS-based mapping application that runs on a mobile GIS device with integrated data collection and sub-foot GPS capability provides field crews with the required current information and allows for complete synchronization with a centralized GIS and database. Integrating operator defined data collection templates with the mobile mapping tool ensures that all parties are collecting information consistently and accurately. Validation of all collected data is programmatic and any non-compliant entries are identified for correction. Once validated, data is integrated into a central platform allowing for broad dissemination of consistent information across the business. Working with data that has been validated and integrated, guarantees a robust, defendable and repeatable analysis. Consistent integrity management program development can now occur to assist in the development of preventative and mitigative (P&M) measures and to ensure decision making is sound. As can be seen in Figure 10, the implementation of field-based maintenance activities allows an opportunity to assess the value from defined integrity programs and begin the data collection process anew.

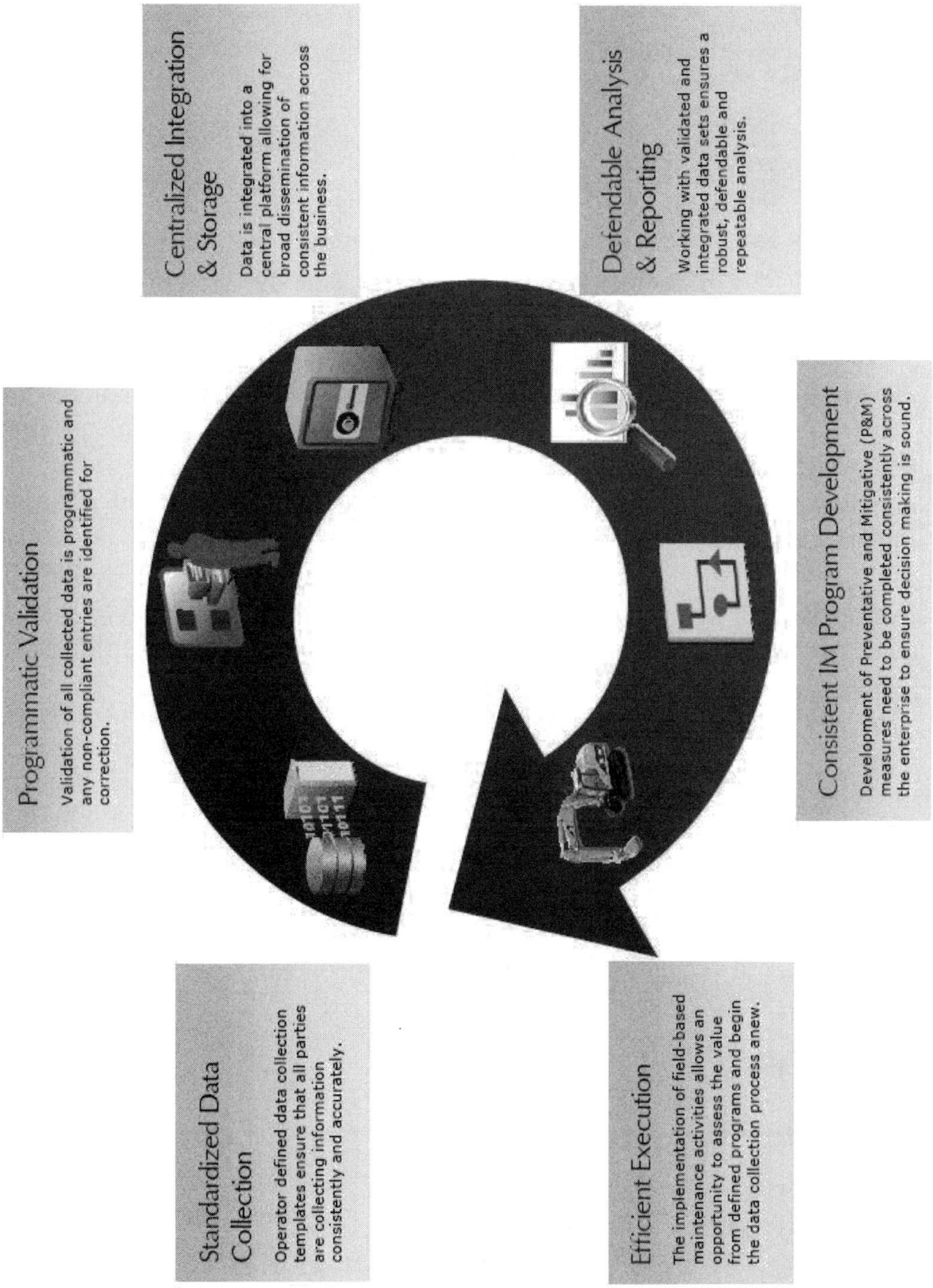

Figure 10 – Integrity Data Management Workflow

By following an integrity data management work flow similar to this, the future will see more pipeline operators taking ownership of their data, as this forms the core of their GIS and affects the validity of any subsequent analysis.

References

(1) American Society of Mechanical Engineers, ASME B31.8S, "Managing System Integrity of Gas Pipelines," 2004

(2) American Petroleum Institute (API) Recommended Practice (RP) 1162. "Public Awareness Programs for Pipeline Operators," First Edition, 2003

(3) American Code of Federal Regulations (CFR), Title 49: Transportation, Part 192 – Transportation of Natural and Other Gas by Pipeline: Minimum Federal Safety Standards, 2002

(4) American Code of Federal Regulations (CFR), Title 49: Transportation, Part 195 – Transportation of Hazardous Liquids by Pipeline, Implementing Integrity Management – Final Rule (as amended), 2007

(5) American Petroleum Institute (API), Standard 1160, "Managing System Integrity for Hazardous Liquid Pipelines," First Edition, 2001

(6) Zuczek, P., Deng, C., Adams, K., and Mihell, J., "An overland-hydrographical spill model and its application to pipeline consequence modeling," Proceedings of the International Pipeline Conference, IPC2008-64389, Calgary, Canada, ASME, 2008

5 Our Environment

5.1 Introduction

A major theme in this book is how geomatic practices and technologies are changing and innovating the way we design, build, and operate pipelines. A section on our environment is similarly important because how we consider our environment in the pipeline lifecycle is also affecting the pipeline industry more now than at any time in the past. Therefore, it is worthwhile to focus on areas of change and share them with the pipeline community so that we may progress our field. We have differentiated this chapter from the previous chapters by focusing less on the technical design of pipelines and more on how the pipeline, during design and once built, will affect and be integrated into the physical landscape and into the lives of people. This chapter will demonstrate how geomatics is able to provide the spatial context that regulators, land use planners, and the public seek when trying to decide the best way to integrate pipelines into our environment. The main issues we will be addressing in this chapter are; taming the uncertain regulatory process, community relations, wildlife and habitat, water, and a look at what's ahead.

Like most areas where geomatics is applied, environmental considerations on pipeline projects benefit from geomatics' ability to answer questions related to specific themes. Through this chapter we will focus on **four central themes** that show the interrelationships between geomatics, pipelines, and our environment. These are:

Theme 1 - Proximity – How close is the pipeline to various flora and fauna, riparian zones, water crossings, residential communities, archaeological sites and any other features of interest near the pipeline that may be affected due to the relative proximity to the infrastructure?

Theme 2 - Data management – How do you deal with terabytes of geo-spatial data with respect to water chemistry, soil samples, air quality, environmental regulatory criteria guidelines, land use management areas and cadastral boundaries. Furthermore, we need to address the fact that these data are always changing?

Theme 3 - Data integration – How do you combine these data sets to make sense of biophysical processes or to model phenomena (e.g. effect of pipeline on vegetation communities, wildlife habitat suitability indices and migration corridors, air emissions, and groundwater modeling) that can be used to model the potential effects of the pipeline on the environment before its actual construction?

Theme 4 - Reporting – How do you communicate your data, findings, and analyses in a manner that clearly illustrates the integration of many aspects of environmental data and the importance of the spatial context of the pipeline development for a variety of stakeholders?. What will the pipeline mean for the environment and for people?

The breadth of environmental impact assessment for pipeline planning, maintenance, and decommissioning is vast (1). Although there are additional and emerging environmental pipeline applications, our four central themes of proximity, data management, data integration, and reporting can be applied widely and possess a generality about them that can be easily implemented in new ways to environmental problems. In each section we look at how the central themes have been applied to pipeline projects and use examples wherever possible to demonstrate practical implementation.

5.2 Taming the Uncertain Regulatory Process

Designing and building pipelines is a technical challenge, an exercise in public relations, and a regulatory administrative epic at times. Although construction costs are a paramount concern on pipeline projects, as engineering and construction methods improve, an uncertain regulatory process is creating the emerging risk profile for many projects. Although regulatory compliance is covered in detail in Chapter 6, we think it is also necessary to mention it as part of the larger environmental framework since it is ultimately the pinnacle of environmental data collection and analysis performed.

What makes the regulatory process seem uncertain rests largely in the unpredictability of the regulatory process; when will we get our permits? When will we be on stream? Criteria and guidelines related to air, water, soil and land use management change over time as new information becomes available and also change over regions from territory to territory. Hence, the answer to these questions are held largely by the regulator, however, the ability to enable the process and react to requests and changes by the regulator lies with the filing organization. Dealing with the uncertainty is done on many levels and through as broad an approach as possible. The best example of this is in keeping up to date with changes in guidelines and being conservative when designing the pipeline in terms of potential environmental impact such that you will meet the guidelines easily even if these criteria change over time. Where geomatics is helping out can be summarized simply as *being as organized as possible to respond to questions and requests in a manner consistent with previous applications and discussions, while retaining any changes or revisions for future engagements with the regulator and project progress*. This is accomplished through data management, integration, and reporting. Let's look at an example.

An organization has filed Environmental Impact Assessments and the regulator has made an information request to see certain landscape features, the location of workspaces, and optional facility sites for a large pipeline. With only 4 weeks to respond, how is this best achieved? Visual maps or alignment sheets would certainly be an effective depiction of this information. As is described in Chapter

3, automating map production or alignment sheet generation would allow the organization to complete such a project in the time allowed. The key to timely completion would contain three parts:

Theme 2 - Data management - The data is organized, centrally managed, and has a change management process to provide integrity. It is from the central data repository that maps, route sheets, alignment sheets, or any other reports are generated. Changes in the data will be reflected in any new report generated from the database. This provides integrity in the output and confidence and consistency in what is being communicated.

Theme 3 - Data integration - The data can be combined and overlaid to accurately fulfill the information request.

Theme 4 - Reporting - The final product can be generated and output in such a way that the first run as well as revisions are simple and fast. This is achieved through auto-mated map and alignment sheet generation, essentially reports which pull data from the database and display it in a standard template.

This level of organization is forming a new reality in dealing with regulators. What's happening is that regulators are getting used to, and coming to expect this level of response. It is not unreasonable to suggest that soon the current level of sophistication will be standard in the regulatory process.

5.3 Community Relations

When determining a pipeline route, we normally look at engineering principles, i.e. determining the features we want to avoid and areas where we would prefer to construct that provides the path of least resistance to develop a cost estimate for this project. However, what we consider our best objectively defined cost-estimate route is not always a best route when communities are consulted. This is an example of what pipeline designers and operators deal with and is telling of how our communities are critical to the general acceptance and progress of large and small pipeline projects and their operation.

There is a critical balance pipeline companies experience when engaging our communities. Involve communities too early in design and planning is overwhelming, engage them too late and the communities feel excluded. The public wants to know about potential impacts and how they will be mitigated. As a result, many pipeline companies are successfully cooperating and collaborating with the public by following some progressive ideals like *transparency, consistency, responsiveness, innovation, and information sharing*.

So how does geomatics help achieve these ideals so that timely permitting and construction will occur while addressing public issues and concerns? All four of our central themes have made an impact on community relations.

An example of this has been a pipeline project where a number of communities including aboriginal groups and land owners have concerns about land access, routing near protected areas, and benefits such as local employment and traditional aboriginal land use. The two main components that make this process as smooth as possible are; staying organized, and communicating clearly. Let's look at how we achieve this with our four central themes.

Theme 1 – Proximity - Using a GIS, the environmental consultant shows where the landowners and protected areas are in relation to the proposed route. As the consultant receives feedback, proximities can be determined from new queries. The proximities to the pipeline will vary depending upon the range of effect that may be of concern. For example, due to the broad distance that air and water can travel the proximities of these to the pipeline may be considered at a much greater distance within a regional study area versus the proximities of animals whose ranges will also vary depending upon size and natural migration behavior but may only extend to a more local study area.

Theme 2 - Data Management - This is a critical factor in staying organized since the data used and generated during design and consultations will continue to change. As a result, maintaining data revisions (like the current route, new information obtained during or after consultation) are critical to providing consistency to the process. Furthermore, an agreement to share and post information is best supported in a GIS where data is stored and maintained.

Theme 3 - Data Integration - To competently tell a story and describe routing and land access scenarios, data needs to be integrated. Data layers are overlaid onto each other to study proximity and perform analyses such as noise effects. In addition, traditional knowledge from the aboriginal communities is used to augment scientific findings from field studies. This integration is becoming more common as we discover ways to combine the two complimentary forms of knowledge.

Theme 4 – **Reporting** - This theme is perhaps one of the greatest contributions made by geomatics to community discussion through the application of innovative cartographic output and 3D digital environments. One of the easiest ways to get a point across is through visual display. This project used color maps with current ortho-photos backdrops to display the proposed route, parcels, landowners, crossings, protected areas, aboriginal traditional use areas, and local water resources. 3D simulations were used to assess line of sight for pipeline facilities such as that shown in Figure 1. Furthermore, proximities are shown when a buffer is drawn around the facility (Figure 2). This illustrates a regulatory footprint which can help designers better place the facility taking into account, line of sight, protected areas, and other features to avoid.

Figure 1. Line of sight analysis to determine if a proposed pipeline facility and right-of-way is visible from a popular campground. The white cross-hatch areas are those parts of the landscape seen directly from the campground when looking towards the proposed route.

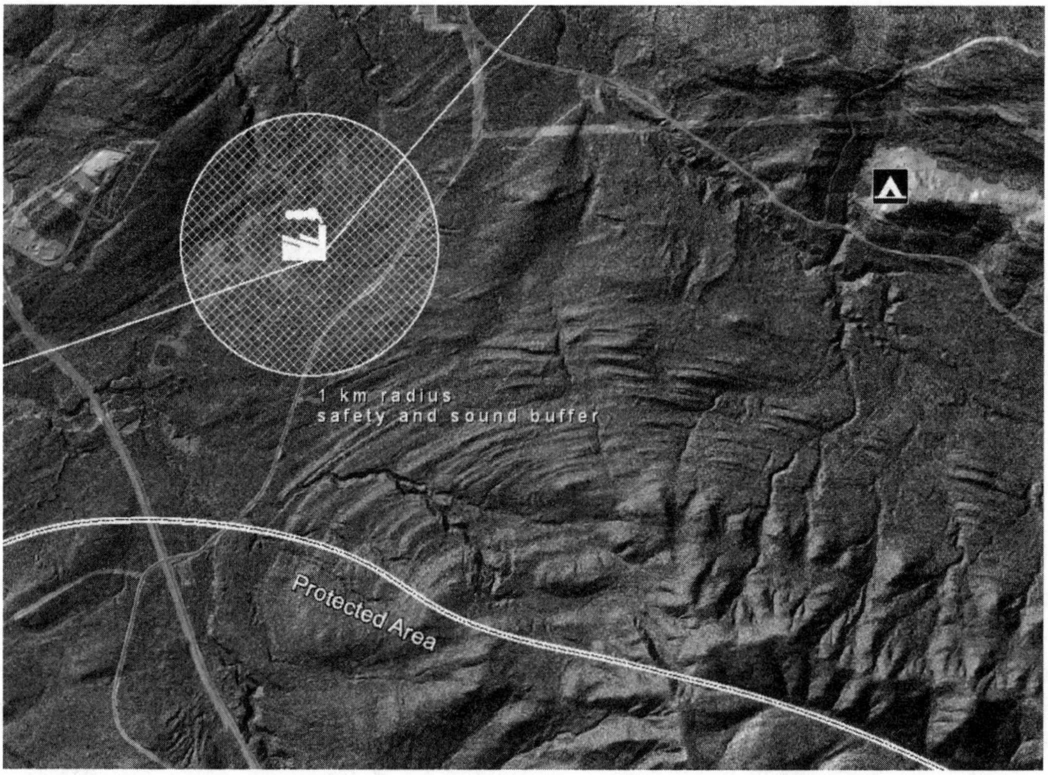

Figure 2. A 1km buffer drawn around the facility to show the sound and safety footprint desired by the community. This helps place the facility in relation to other features that must be avoided (e.g. protected areas).

Combining these geomatics technologies with other tools and information such as Issues Management Systems and traditional knowledge is a more inclusive and information-rich process. Managing and mapping the spatial and non-spatial information in a transparent straightforward manner so that various stakeholders can understand in laymen terms sets a stage where involving the public ultimately becomes collaborative and less contentious.

5.4 Wildlife and Habitat

Some of the higher profile environmental assessments occur in the realm of wildlife and habitat. Geomatics, and particularly GIS, has had a long history of use in environmental assessment within the pipeline industry. This is largely because wildlife and habitat are a are very dynamic spatial phenomena with characteristics that are effectively managed and maintained using GIS. Species distribution, habitat features, disturbance mapping, modeling effects, among many other issues are regularly dealt with using GIS. Perhaps the best way to illustrate geomatic applications to environmental issues is to delineate how our four themes influence *Baseline* Information and *Environmental Impacts.*

5.4.1 Geomatics and Baseline Information

Baseline information is essentially the starting state of the environment before pipeline or facility construction. It provides the environmental datum which future impacts can be relatively compared to. Collecting baseline information is a fundamental part of the Environmental Impact Assessment (EIA) process and can result in volumes of data and information that must be collected, managed, and analyzed through the EIA workflow. The International Association for Impact Assessment (IAIA) defines an EIA as

The process of identifying, predicting, evaluating and mitigating the biophysical, social, and other relevant effects of development proposals prior to major decisions being taken and commitments made (2)

The main benefits of geomatics and particularly GIS for the collection of baseline information are seen in compiling species distribution data and habitat features. Our central themes that best service baseline information are Data Management and Reporting. Although not a central theme throughout this chapter, we feel it is important to highlight the application of digital data capture through mobile computing since field data collection capabilities are rapidly growing and having a strong impact on field programs.

Theme 2 - Data Management - Collecting and storing environment related project data can be an immense task. By using a GIS to store and manage data, one can deal with large disk volumes, features and attributes (e.g. water chemistry, soil samples, vegetation inventory), and even multiple scales of data (e.g. local vs. regional study areas). Effective data management allows teams to manage change and versioning and compile metadata for historical record.

Theme 4 – Reporting - When data has been compiled, it typically needs to be output in some fashion in order to communicate the distribution of features and phenomena. This is done through cartographic output or formatted reports such as those created from a file review process.

5.4.2 Data Capture from Mobile Computing

Mobile computing has seen some significant application in recent times as mobile units (e.g. tablet PC's, GPS units) and web-based applications have become more ubiquitous in supporting dynamic field collection (3). This technology can be used to collect spatial information and its attributes throughout the entire lifecycle of environmental field data collection: Pre-field, Field, and Post-Field. You can find an interesting compilation of case studies, articles, and Podcasts on this subject (4).

Pre-field - Essentially, a person in the field has a mobile computing device that is enabled with a GIS and some pre-loaded data for navigation and orientation.

Before the field survey even begins proposed sample locations, ecological study area quadrants, and previously collected information can be assessed and uploaded to make the field work more efficient and meaningful.

Field and Post-field - During the actual field work data is input directly into the device and can be transmitted directly back to a central database in real-time or uploaded at a later time. The benefit here is significant, field workers can immediately see the spatial distribution of field data on the mobile computing device as they collect the data and data is only collected once in the format it will be stored in and analyzed.

Examples of this are seen in the updated methods of data collection executed by vegetation, wildlife and aquatics biologists in using mobile computing for their pre-field preparations, field executions and post-field data compilation. Before going into the field useful data such as environmental data (e.g. previously surveyed rare plant locations, rare ecotype locations), useful base features for navigation and access concerns (e.g. hydrology, roads, cut-lines) and pre-field points of interest to visit can be uploaded to improve the quality of the field survey. Once out in the field sample points, field survey and observation locations can be collected including any attribute information that would be of interest to the biologists at the time of survey (e.g. species type, sample location id, sample depth, water condition, etc.). Then after the field collection this data can be uploaded into the data management system for further data integration and reporting.

5.4.3 Geomatics and Environmental Impacts

The question of how a pipeline and its facilities will impact the environment is addressed through the development of analyses based on comparison to the baseline information. Again, GIS is the best tool to deal with this process because it fully supports scientific analyses and algorithms, showing the geographic extent of effects, and can be used to understand the duration and magnitude of an impact. Due to the very nature and breadth of determining impacts, all four of our themes would apply.

Theme 1 – Proximity – By spatially enabling the data collected and stored, we can determine how close features are to each other and therefore look deeper into their effects. If noise from a compressor station affects bird nesting, then it is important to know how close nesting areas are. Creating a right-of-way may alter the size of wildlife core security or refuge areas.

A recent study was conducted to determine core security areas (CSAs) for a signature species based upon the effect of development. Analysis uses buffers, proximity analysis, and spatial modeling to determine the cumulative effects stage in time where all past, present, and future footprints may be implemented within the larger regional study area. A buffer has been placed around any anthropogenic developments to determine the potential fragmentation of the study area into varying sizes of core security areas where the species may find potentially suitable habitat. The study ultimately shows the habitat suitability index (HSI) which has been modeled and mapped, displaying the areas that will be of least and greatest value to the species population within the area of the development (5).

Theme 2 - Data Management - Not only is it important to store and manage volumes of baseline information, but to extend those efficiencies to impact analyses, so results may also be stored and managed. After all, results (i.e. impacts) are also data. The most effective means of data management is by trying to store as much information spatially and centrally as possible. Given the large amounts of information that can be generated, it is important to have a sound data management system and change management process to do this.

Theme 3 - Data Integration - GIS gives us the ability to powerfully model phenomena and do what-if scenarios in a spatially distributed way based on the data we have stored. The study discussed above comparing the effects of varying pipeline footprints on the core security areas or habitats of varying wildlife species provides a good example of this. Different species will respond differently to the development. The effects of burns or potential burns could also be considered for changes to habitat and species due to these somewhat stochastic events.

Collecting baseline information and considering impacts also leads into the closure and reclamation process which is also an essential part of the regulatory process. Data integration can also be used for this part of the regulatory process. By determining the potential environmental effects – not only can mitigation of such effects be planned before the actual construction of the pipeline but various reclamation planning can be modeled out and be ready to be put into motion for the future closure of the pipeline once the life of the pipeline is finished. The beauty of all this is it can be modeled in scenarios over and over again, all the while, revising data depending on the results.

Theme 4 – Reporting - To effectively communicate these results, there has to be some compelling output. GIS has immense flexibility and elegance in outputting information. Environmental output can be in the form of maps and drawings as well as automated alignment sheets. The latter are particularly powerful since they can be generated quickly and illustrate anything stored in the database that is related to the route. An excellent example of a GIS based environmental impact assessment available for public viewing is the one created for the massive Mackenzie Gas Pipeline Project (6). In addition, the renewal EIS for the Trans-Alaska Pipeline System (TAPS) is available on line for the public to see (7).

GIS has enabled pipeline designers and operators to collect, understand, and communicate data and information on potential impacts, the phases where they may occur (e.g. construction, operations, decommissioning and abandonment). These technologies and methods are transforming the pipeline and environmental industries by enabling better data collection, storage, management and communication in an inherently open and public process.

5.5 Water

We all care about water and our concern for its abundance and quality make it potentially a very contentious issue for pipeline designers and operators. This places a lot of pressure on environmental professionals and design engineers to carefully and accurately understand and respond to any impacts to our water, specifically its quality and availability. Furthermore, water is needed in the pipeline process for things like directional drilling and hydro- testing, so as a result, water becomes a valuable resource to the pipeline industry as well. Similar to wildlife and habitat, water is typically treated in the environmental review and impact process.

Environmental water quality and quantity is considered from a groundwater (below ground) and surface water (above ground) perspective. Groundwater can be usefully extracted from underground aquifers using water wells and surface water can also be used for agricultural, industrial and human consumption. Surface water bodies are also used for recreation and serve as riparian habitats for aquatic wildlife. These riparian habitats include areas where pipelines are concerned with water crossings including streams, rivers, lakes, ponds and wetlands. Given the dynamic hydrologic cycle and the sensitive nature of some of these habitats to disturbance, mapping, visualizing and integrating data for all involved is useful. The four themes assist with:

Theme 1 - Proximity - Groundwater and surface water bodies can be vulnerable to overuse and to the improper use and disposal of chemicals and fracking muds (from directional drills) and can be difficult to reclaim to its original state. GIS can be used to provide a spatial analysis of the proximity of the pipeline to any vulnerable water areas.

Theme 2 - Data Management - All of the data required to model and map water resources comes from many different sources including geological information, water chemistry, water quantity, and pipeline information. The collection of this assortment of data can result in significantly large data sets that can be challenging to organize into a coherent structure such as a data model (Chapter 7). This complexity in data structure lends itself well to GIS and databases to effectively manage the data sets centrally and allow for efficient data access and querying of the data sets.

Theme 3 - Data Integration - GIS allows for diverse data sets to be integrated together for determining the relative vulnerability of water resources to pipeline development using a combination of groundwater data, geological information, topography and anthropogenic footprints. Once data is integrated spatial analysis can begin and an assessment can be performed developing an inventory of baseline information and modeling potential impacts.

A good example of this has been the use of the USA's Environmental Protection Agency's (EPA) DRASTIC groundwater vulnerability model. This is a methodology developed by the EPA that allows the pollution potential of any hydrogeologic setting to be systematically evaluated (8). Successful models such as this dynamic model spatially integrates information related to soils, topography, aquifer media, hydrology and other potential environmental risks using GIS to produce a weighted relative groundwater vulnerability rating.

Theme 4 - Reporting - This is where GIS can be of the greatest benefit in providing clear visualizations of features of interests. This is especially true of the groundwater component since it not only needs to be visualized in a 2-dimensional way as surface water bodies can, but also includes the additional dimension of depth below ground. Continual advances in computer hardware and 3-dimensional visualization can assist in bringing visual clarity to explaining the subsurface or hydrogeology of an area with cross sections and cut away views. The ability to visualize this sort of information is valuable in explaining complex data sets to a variety of stakeholders including regulators and the public.

5.6 *What's Ahead?*

Throughout this book our goal is to demonstrate how geomatics is helping to change the way we build and operate pipelines. We see this happening through the innovation introduced by technologies and processes. For a discipline to be successful it must be able to change and evolve and we feel geomatics is supporting this evolution. With each day and each new pipeline permit application, our environment is coming under greater pressure as are producers and pipeline companies to comply with regulations and public perception. This motivates us to ask the question: What larger environmental momentum is taking shape? What are the hot environmental topics for the near future that geomatics is helping pipeliners deal with? Here we look at three; *population and development growth, demand for environmental enforcement, and cumulative environmental effects*.

5.6.1 Population and Development Growth

In North America, we don't always think about running out of space for oil and gas wells, pipelines, and facilities. However, encroachment is a very real problem since all the features that surround us, cities, water, industrial development, roads, utilities, protected areas, even our own air are all dynamic and part of a network of everything else. Therefore, we need an effective way to deal with what is out there, where it is, and how it is changing.

Fortunately, GIS has some strong capabilities in this area and future trends show some growing potential as well. The growth issue rests ultimately with regulators since they issue the permits, however, since everything is connected to everything else, it would be naïve to place all the responsibility with regulators. The technologies that would help in this area are proximity and networks using GIS, centralized data management, and the GeoWeb.

We have already discussed proximity calculations using GIS, where one can relate something to something else based on where it is. Since all data in a GIS is spatial, performing analyses in a spatially distributed way is easy. Furthermore, when we know where everything is, we can also deal with how everything that is connected to each other will affect each other when changes occur. This is the function of GIS networks. These are not networks like computer or digital communication networks (although based on similar principals) these are networks formed between data. For example, a pipeline system is an obvious network and the features that cross it (like other pipelines, roads, rivers) or intersect it (HCA's, protected areas) are also part of this network. When something changes within the network, like a new pipeline is crossed, new houses are encroaching, or protected areas are expanded the effect on the entire network can be determined. Environmental impacts can be updated, new alignment sheets can be generated, and emergency notification systems are updated. Also important to consider are the nodes in the network, such as compressor stations, that will feel the change in the system since more or less capacity is added or new parts are added to the network.

Centralizing data would allow regulators to have a single clearing house of information which they can post and update for organizations and citizens. This is already happening with the U.S. Office of Pipeline Safety (OPS) and the National Pipeline Mapping System (NPMS) which essentially stores and shows the location of gas transmission and hazardous liquids pipelines, liquefied natural gas (LNG) facilities and breakout tanks within the command of OPS. What gives this strong momentum to continue and expand into the future is growing internet access and the availability of information through the internet. Given that information on the internet ranges from the factually reliable to the completely insane, it would be in an operator's and regulator's best interest to post their own accurate information about the status of a pipeline or project rather than have someone else's interpretation be the only content available. This is best supported by central data management since it keeps the story straight by broadcasting on one version of information, thus reducing confusion among the public and other stakeholders.

The next technological leap in the data management and sharing world would be the GeoWeb. This is fully explained in Chapter 8, but to briefly summarize, it can essentially be considered a global peer-to-peer GIS, where spatially enabled data is served up on the world's largest GIS framework. Just like we use peer-to-peer networks to share digital files, we will do so for spatial data that could be served up from home computers or large servers depending on the data set. The Nature Conservancy is undergoing pilot studies using the GeoWeb to link diverse and disparate datasets for ecosystems around the world (9). This puts data creation, hosting, and dissemination in the hands of everyone, and not exclusively held by large organizations. A recent book on the GeoWeb has several examples of environmental applications including natural hazards, air quality, and disaster and crisis management (10).

5.6.2 Demand for environmental monitoring and enforcement

Governments in North America and all over the world are continuously strengthening their environmental legislation which pipeline companies must comply with to build and operate. Although these laws and guidelines are to serve the interests of citizens and future generations, they can be confusing and difficult to enforce. With government departments under-resourced to monitor and enforce legislation many non-governmental organizations (NGOs) and communities are still able to use these laws to challenge or question projects. This is very good thing when the public has tools at its disposal to act in its best interest. In fact, with the growth of internet accessibility, people have more access to data and means by which to collaborate than ever before.

As a result, the public has a growing involvement in projects that should not be overlooked, and as mentioned in the last section, with land availability shrinking, demand for energy increasing, and encroachment growing, cooperation on all sides is the best path forward. The last topic showed how geomatics was supporting government and the public with data centralization and soon the GeoWeb. These technologies hold true for this issue as well. The better regulators or any other agency can keep track of what is going on, the better they can mobilize resources to monitor and enforce legislation. As we stand only at the beginning of the information age, the ability of people, governments, and other organizations to mobilize, cooperate, and communicate information and opinions is an unavoidable reality that we must prepare for. In addition, as the current managerial generation begins to retire, the potential loss of knowledge and its impact may be felt if there aren't effective information and data management systems to retain knowledge and business processes. The growing power of GIS and spatial data management is one part or the solution to preserve and share information, which is proving to be as valuable an asset as the pipeline itself.

5.6.3 Cumulative environmental effects

A subject that is vital to addressing environmental effects and sustainability of our pipelines is cumulative effects. Many regulators are now understanding and realizing that it is important to model the cumulative effects that may occur as a result of a network of many developments in any given area. For example a pipeline that is in the vicinity of a Brownfield site or urban area will have many different risks and cumulative environmental impacts over a green field site. This concept has been summarized by the Canadian Environmental Assessment Agency (11) as:

The effect on the environment which results from effects of a project when combined with those of other past, existing and imminent projects and activities. These may occur over a certain period of time and distance.

What makes this an obvious and significant topic for our industry are that three main components of cumulative effects are present; progress of time (inescapable part of our universe), more projects putting more pipe in the ground, and all these assets take up space while there is less room to put all this stuff. How is geomatics helping this issue? Since the essential nature of geomatics is to deal with spatial phenomena, cumulative effects can be effectively handled. In fact, a one of the significant advantages of a spatial data management system is its ability the maintain versions of data. Therefore, we are able to efficiently accumulate data on environmental changes (i.e. time-based data) over geographic areas (spatially-based data). This combination of time and space makes GIS and important tool in compiling, analyzing, and reporting cumulative environmental effects. Figure 3 illustrates how geographic and temporal baseline information can be gathered and then how past, present, proposed projects and possible future actions (i.e. developments) combine to produce cumulative effects that need to be assessed using impact assessment methodology (i.e. overlay mapping and GIS).

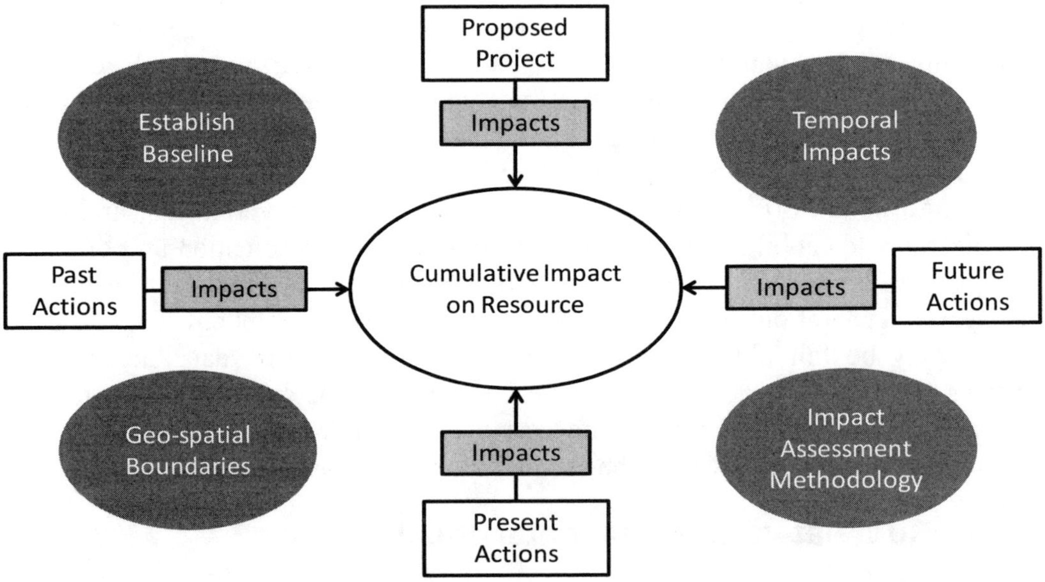

Figure 3 - Conceptual cumulative effects model, considering both temporal and spatial aspects of a project

Over time and with the development of many visualization and modeling software packages environmental regulators may in the future consider requesting the data in a format that they are able to upload and use for their own internal analyses with respect to permit applications. The natural outcome of this would be the request for spatial data by environmental regulators of permit applicants. This data could then either be compiled for future public use or for use within government itself. Given the necessity for permit applicants or environmental regulators to provide evidence of their due diligence in the permitting process with respect to proving the soundness of the data used in the regulatory process it would be prudent for anyone performing cumulative effects assessments to consider establishing robust data standards and guidelines for collecting and obtaining environmental regulatory information. A robust data set that has proper metadata on source, scale, accuracy, and precision is information that can be stored, integrated and migrated easily. Organized coherent data sets can greatly assist environmental regulators and permit applicants using the data sets to be better able to efficiently make informed decisions with respect to allowing development that meets environment and safety standards and where the permit applicant has considered public and socio-economic concerns of any stakeholders with respect to the pipeline development.

With cumulative effects the most effective way to analyze the potential effects of multiple developments is with GIS and databases as it allows for the integration of many diverse data sets over varying spatial extents. Cumulative effects involves considering information gathered from a great number of sources and so the integration using GIS is effective and necessary (e.g. overlay analysis of many different development footprints within the vicinity of the pipeline combined with information regarding sensitive areas). GIS offers this opportunity to store and query data in a way that can be effectively analyzed for environmental regulation. Spatial data integration also provides the spatial context that is necessary in explaining the effects of environmental impacts so clearly in an immediate and intuitive manner to all – land use planners, engineers, regulators and the public.

Given the advances with respect to the GeoWeb – information can be displayed in a way that can be served to all from a central source to many making this an effective way to serve out data. More and more people look towards this form of communication as it is rapid, immediate, and can be updated efficiently through a single source.

With respect to the environment and pipeline permitting it is essential for anyone involved in the process who wants to be innovative in their approach to the environment to consider geomatics as having effective tools that can assist in the regulatory process. By having robust coherent centralized data sets that can be trusted, effective data integration and using the reporting tools to clearly demonstrate potential effects of the pipeline routing on the environment it allows trust to be built up between the permit applicant, the regulators, the public and any other stakeholders who may be involved in the application. This is the basis of environmental process which is the desire to achieve as close as possible the progressive ideals that were discussed earlier in this chapter in community relations like *transparency, consistency, responsiveness, innovation, and information sharing.*

References

(1) Kerr, D., 2004, "Environmental Impact Assessment of Large Pipeline Projects in Remote Environments: Integrating Engineering and Environmental Design", Proceedings: Terrain and GeoHazard Challenges Facing Onshore Oil and Gas Pipelines, London, UK., 2004

(2) International Association for Impact Assessment in cooperation with Institute of Environmental Assessment, UK. 1999. Principles of Environmental Impact Assessment Best Practice. And also through http://www.iaia.org/modx/

(3) Belussi, A., Catania B., Clementini, E., and Ferrari, E, "Tracking of Moving Objects with Accuracy Guarantees", Springer Berlin Heidelberg, New York, 2007

(4) Refer to http://www.esri.com/software/arcgis/arcgismobile/success_stories.html

(5) Refer to http://www.petro-canada.ca/pdfs/VolumeIICSection4-Figures48-59.pdf

(6) Refer to http://www.mackenziegasproject.com/theProject/regulatoryProcess/applicationSubmission/Applicationscope/EIS.html

(7) Refer to http://www.tapseis.anl.gov/

(8) Refer to http://nepis.epa.gov/Exe/ZyPURL.cgi?Dockey=20007KU4.txt

(9) Refer to http://www.nature.org

(10) Scharl, A., Tochtermann, K., "The Geospatial Web: How Geobrowsers, Social Software and the Web 2.0 are Shaping the Network Society" (Advanced Information and Knowledge Processing), Springer-Verlag, London, 2007

(11) Hegmann, G., C. Cocklin, R. Creasey, S. Dupuis, A. Kennedy, L. Kingsley, W. Ross, H. Spaling and D. Stalker. 1999. "Cumulative Effects Assessment Practitioners Guide". Prepared by AXYS Environmental Consulting Ltd. and the CEA Working Group for the Canadian Environmental Assessment Agency, Hull, Quebec. And also through http://www.ceaa.gc.ca/

Acknowledgements

The authors of this chapter gratefully acknowledge the ideas and direction provided by Karen Etherington of TransCanada Pipelines.

6. Environmental and Regulatory Compliance

Issues of safety and environmental health are of primary concern to pipeline companies, regulatory bodies, stakeholder groups and the general public alike. As part of day-to-day operations, pipeline owners and operators are required to document an ever growing amount of information to demonstrate their compliance with engineering standards, safety and environmental regulations, and record management requirements set out by various regulatory bodies. In this chapter we will examine a variety of ways in which Geographic Information Systems (GIS) technology is being used – and could be used in the future – in the Pipeline Regulatory Process.

6.1 Using GIS to Support Regulatory Compliance

The very nature of GIS lends itself to supporting many types of regulatory requirements. GIS technology enables users to tie a myriad of information to any given location (1). This ability allows pipeline operators to maintain detailed records, in the form of automated maps and related databases, which store the location and attributes of their assets, and provide that information to the appropriate regulatory agencies.

The value of GIS to pipeline automation is twofold. First, GIS provides an automated environment for performing analysis functions such as risk assessment or defining High Consequence Areas (HCAs). Second, GIS offers the ability to efficiently manage spatial data so that it can be easily recalled at any time, in response to regulatory requirements, information requests, and inspection or audit requisites. In this section we will examine in greater depth some of the most common ways in which GIS can be applied to support regulatory compliance and reporting.

6.1.1 Permit Applications

Before a pipeline is ever approved for construction and operation, pipeline companies may be required to submit several permit applications to indicate the proposed route of the pipeline and where the pipeline may cross watercourses or transportation routes, or for any facility that is proposed along the pipeline route. Maps form a key component of the permit application, and are required to identify the exact location of the proposed infrastructure vis-à-vis the surrounding environment and geological features. GIS is a necessary tool in developing permit application maps for regulatory approval. GIS technology allows a variety of 'what-if' scenarios to be tested and visualized in a matter of minutes, allowing for various design options to be considered and evaluated quickly.

6.1.2 Risk Analysis and Geohazard Identification

As part of the approvals process, most regulatory bodies require extensive documentation reporting on the findings of risk analysis and the identification of potential geohazards. Geohazard identification involves the location, identification, and comparative assessment of historical failure incidents caused by geohazards. A growing movement towards proactive management has resulted in emphasis on terrain analysis using stereo aerial photographs and satellite imagery to map areas of potential influence surrounding the proposed pipeline, beyond the right-of-way and including surrounding watersheds and populated areas. GIS tools are used to support: frequency analysis; consequence analysis; and risk estimation, evaluation, control, and monitoring activities. Pipeline companies are required to provide detailed risk assessment reports to regulatory bodies, to ensure they are compliant with safety protocols.

6.1.3 Responding to Information Requests

Regulatory agencies may require, at any time, that pipeline operators provide information regarding any aspect of their proposed or operating pipeline and/or facilities. Spatial data management, facilitated through a GIS environment, ensures that information on assets, potential environmental impacts, and safety concerns can be immediately collected for any spatial area. This eases the task of responding to anticipated and unexpected regulatory information requests in a timely fashion.

6.1.4 Pipeline Integrity and HCA Identification

Subsequent sections of this chapter will demonstrate how several regulatory agencies – most notably the Pipeline and Hazardous Material Safety Administration Office of Pipeline Safety (U.S) and the National Energy Board (Canada) enforce regulations related to pipeline integrity. The geoprocessing functionalities of GIS allows for the simple overlay of data layers, and proximity/buffer calculations to identify potentially hazardous parts of the pipeline network. Complex algorithms can be built to model the pipeline, surrounding infrastructure and natural features, and potential impacts on pipeline integrity.

6.1.5 Public Awareness Requirements

Public awareness and education programs are critical to minimizing threats to the integrity of a pipeline and ensuring a high level of public well-being. Safety regulations are incorporating requirements for public consultation and awareness programs, including the identification of target audiences, specific messages to be delivered, and the frequency and methods of delivery (2). The American Petroleum Institute's (API) Recommended Practice 1162 provides clarity on how to define target audiences for public awareness programs, which include: the affected public, emergency officials, and local public officials and excavators. While public information resources make it relatively easy to identify emergency officials, public officials and excavators, identifying 'the affected public' can be more difficult.

GIS technology can be used to visualize the pipeline right-of-way with surrounding landowner information, and through proximity analysis can easily identify those properties within the affected corridor. Leveraging the database capabilities of GIS, mailing lists and phone lists can be automatically generated to support mailing programs or emergency contact systems that provide homeowners, tenants, and businesses with public awareness materials or emergency action information. GIS also offers the ability for pipeline operators to leverage their normal business practices to demonstrate their compliance with public awareness and education requirements. GIS can be used to capture and visualize to auditors the scope of the identified affected public, in addition to the type and frequency of the notification provided.

6.1.6 Compliance Reporting and Records Maintenance

Regardless of the governing agency or jurisdiction, pipeline operators will continue to face increasing reporting requirements. Though the scope of reporting obligations is vast, documentation is typically required to demonstrate compliance with engineering standards, safety practices, and environmental rules. With the increasing demands for archival and historic tracking of assets for compliance purposes, the volume of data required to operate and manage a pipeline will continue to grow into the future (3). GIS technology and spatial data management techniques provide a systematic and disciplined approach to ensuring records are well-maintained and readily available to any requesting agency.

6.2 Regulatory Bodies and Legislation

6.2.1 United States

In the United States there are several governmental and independent regulating bodies whose mandates focus on ensuring national pipelines are designed, constructed, operated and decommissioned with minimal risk and impact on humans and the environment. Here we examine the major regulating agencies within the United States and some of the most common regulatory requirements imposed on pipeline companies.

6.2.1.1 Federal Energy Regulation Commission

The Federal Energy Regulatory Commission (FERC; http://www.ferc.gov/) regulates and oversees energy industries in the economic, environmental, and safety interests of the American public. FERC is an independent agency that regulates the interstate transmission of electricity, natural gas, and oil. FERC also reviews proposals to build liquefied natural gas (LNG) terminals and interstate natural gas pipelines as well as licensing hydropower projects. One of FERC's top priorities is to ensure environmentally safe infrastructure. Under this mandate FERC is responsible for: regulating the transmission of natural gas and oil by pipeline in interstate commerce; approving the siting of and abandonment of interstate pipeline, storage and liquefied natural gas facilities; and using civil penalties against energy organizations and individuals to ensure compliance with FERC regulations.

FERC performs operational inspections and audits of industry participants to ensure compliance with rules, orders, regulations, and statutes. Inspections and audits of electric power, natural gas and oil pipeline industries concentrate on materially relevant issues, and enforce penalties for non-compliance.

6.2.1.2 Pipeline and Hazardous Material Safety Administration

Under the jurisdiction of the U.S. Department of Transportation, the Pipeline and Hazardous Material Safety Administration (PHMSA; www.phmsa.dot.gov) administers the national regulatory program for ensuring the safe transportation of natural gas, petroleum, and other hazardous materials by pipeline to industry and consumers. Through PHMSA, the Department develops and enforces regulations for the safe, reliable, and environmentally sound operation of the nation's 2.3 million mile pipeline transportation system. PHMSA oversees the nation's pipeline infrastructure, which accounts for 64 percent of the energy commodities consumed in the United States. PHMSA conducts inspections and audits of pipelines to ensure compliance with safety and training requirements. The PHMSA audits include a review of the accuracy of mapping and survey information for the purpose of identifying the precise location of High Consequence Areas (HCA's) for hazardous liquid pipelines. Operators must identify all pipeline segments that fall within, or could affect, a HCA. Pipeline operators who fail to provide accurate and sufficient documentation to demonstrate their compliance may be subject to significant fines and other disciplinary action.

PHMSA uses GIS primarily for asset management and for decision support when considering new regulations. When PHMSA's inspectors go into the field, they use PHMSA's GIS to locate pipelines and verify attributes. The GIS is also used for emergency response in the event of a pipeline release by providing maps and statistics about the affected area, and contact information for pipeline operators who may be impacted.

Acting through the Office of Pipeline Safety (OPS), PHMSA administers the Department's national regulatory program to assure the safe transportation of natural gas, petroleum, and other hazardous materials by pipeline. OPS develops regulations and other approaches to risk management to assure safety in design, construction, testing, operation, maintenance, and emergency response of pipeline facilities. Since 1986, the entire pipeline safety program has been funded by a user fee assessed on a per-mile basis on each pipeline operator OPS regulates.

OPS is responsible for ensuring the safe, reliable, and environmentally sound operation of the U.S. pipeline transportation system through mandated regulatory and enforcement activities. This mandate is enforced through a number of initiatives including: Compliance Safety; the National Pipeline Mapping System; the Integrity Management Program; Pipeline Safety Data Analysis; Regulatory Development; and, the identification of Unusually Sensitive Areas.

These federal pipeline safety regulations:

- Assure safety in design, construction, inspection, testing, operation, and maintenance of pipeline facilities;
- Set out parameters for administering the pipeline safety program;
- Incorporate processes and rule-making for integrity management; and,
- Delineate requirements for onshore oil pipeline emergency response plans.

OPS has regulatory oversight of approximately 330,000 miles of gas transmission pipeline and 160,000 miles of hazardous liquid pipeline operating in onshore and offshore territories of the US (4). The Pipeline Safety Act, adopted by congress in 1992, directs that OPS must require pipeline operators to identify facilities located in environmentally unusually sensitive areas, to maintain maps and records detailing that information, and to provide those maps and records to federal and state agencies upon request. To store and manage this location information, OPS implemented the National Pipeline Mapping System (NPMS).

The NPMS is a GIS and consists of geospatial data, attribute data, public contact information, and metadata pertaining to the interstate and intrastate gas and hazardous liquid transmission pipelines, liquefied natural gas (LNG) plants, and hazardous liquid breakout tanks jurisdictional to PHMSA. The NPMS does not contain information on interconnects, pump and compressor stations, valves, direction of flow, capacity, throughput, or operating pressure. In addition, distribution and gathering pipelines are not included in the NPMS.

The NPMS is built from data submitted by pipeline, LNG plant, and breakout tank facility operators. In 2002 participation in the NPMS was no longer voluntary for pipeline operators; congress mandated participation through an amendment to the Pipeline Safety Act. Section 15 of the Act details new requirements for pipeline operators, with specific regard to spatial information. The Act dictates that pipeline operators are now required to submit "geospatial data appropriate for use in the NPMS or data in a format that can be readily converted to geospatial data." The revised Act also requires that attribute data and metadata for all pipeline operation systems be submitted for inclusion in the NPMS. Transmission pipeline and LNG plant facility operators are also required to update their submissions on an annual basis. Updates must reflect any and all changes in geospatial data; attribute data, metadata, or public contact information. Updates are required one year from the date of an operator's most recent submission. Breakout tank operators are able to submit data to the NPMS on a voluntary basis.

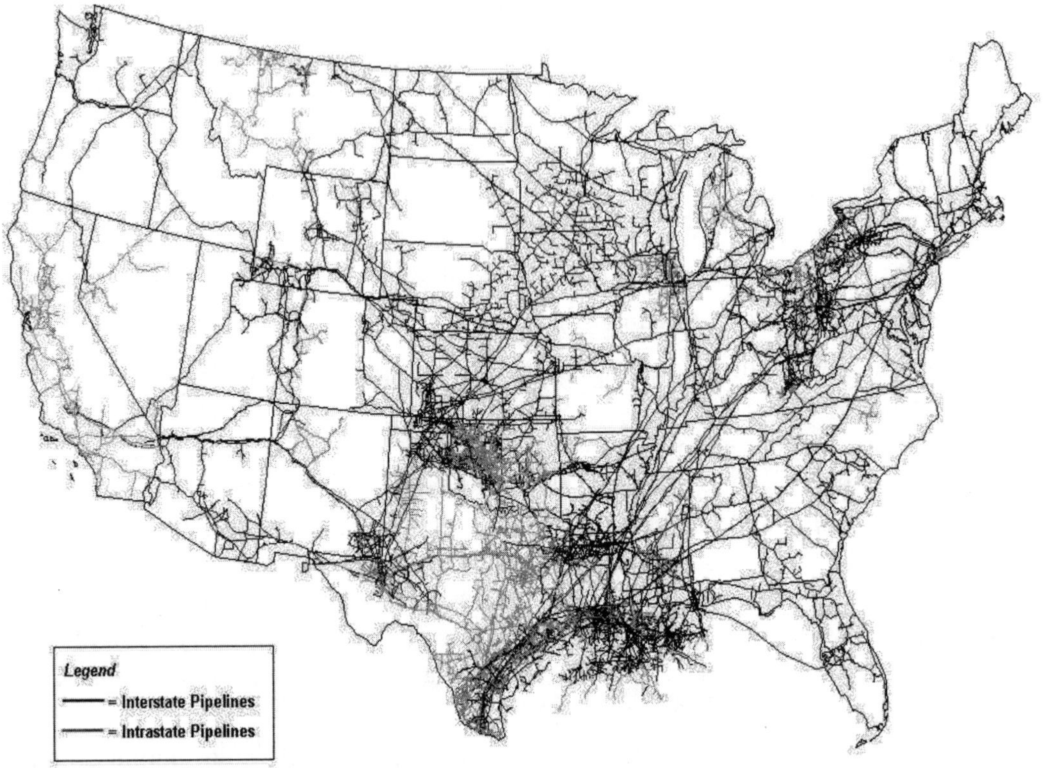

Figure 1: Major Gas Pipeline Systems in the United States. Source: Energy Information Administration, Office of Oil and Gas, Natural Gas Division, Gas Transportation Information System

The NPMS consists of a single National Repository that processes all NPMS submissions. The National Repository serves as the final processing and storage facility for all pipeline data. The National Repository produces a random sampling of check plots after incorporating the operator's geospatial data and attribute data into the repository. These check plots are returned to the operator for review as part of the quality control process. All data passes through a final series of quality control checks before the data are made available to users. The data will be made available to pipeline operators and Federal, state, and local government officials.

Various state agencies currently request or require that operators submit pipeline and LNG data to them. Some state agencies are using the operators' data to create a digital pipeline and LNG layer for their state. NPMS does not supersede or replace state regulations. Operators must still comply with all applicable state regulations.

First developed as a national repository, the NPMS now serves as a decision-support tool for inspection planning, community access and risk assessment identifying where additional precautions are required to guard against potential pipeline releases.

In 2001, after 9/11, the NPMS was removed from the public domain. This was done to protect the security of the pipeline infrastructure. The Public Map Viewer was released to the public on April 2, 2007. This new application allows the general public to view maps of transmission pipelines, LNG facilities, and breakout tanks one selected county at a time.

6.2.2 Canada

Similar to the United States, Canada also has a number of regulating bodies that govern the design, construction, operation and decommissioning of pipelines.

6.2.2.1 National Energy Board

The National Energy Board (NEB, www.neb-one.gc.ca) introduced a GIS program in 2003. Over five years the use of GIS has evolved from simple tasks, such as locating a pipeline and visually assessing its vicinity, to more sophisticated consequence analyses in support of the NEB's Risk-based Life Cycle approach to regulation. GIS usage now expands to the entire lifecycle of a pipeline, from the moment a pipeline application is considered by the NEB to the day the pipeline is no longer needed.

6.2.2.2 The Role of NEB in Governance of Pipelines in Canada

The National Energy Board (NEB) is an independent Canadian federal agency that regulates international and interprovincial aspects of Canada's oil, gas and electric utility industries, including pipelines (5). The NEB promotes safety and security, environmental protection and efficient energy infrastructure and markets in the Canadian public interests. In 2009 the NEB regulated about 70,000 kilometers of pipelines across Canada, shipping crude oil, petroleum products, natural gas liquids, and natural gas (Figure 2).

In 2008 45,000 km of these pipelines shipped over $127 billion (6) worth of products equivalent to 10 per cent of Canada's gross domestic product. An estimated cost of transporting these products was $4.4 billion. This demonstrates huge economical value that pipeline systems bring to the country.

Like the PHMSA (Pipeline and Hazardous Material Safety Administration) in the USA both organizations regulate pipelines. However, several differences exist in the way agencies conduct their regulatory oversight. The NEB also has provincial counterpart agencies in the provinces of Alberta and British Columbia where significant oil and gas activities take place. The Energy Resources Conservation Board of Alberta (ERCB) and British Columbia Oil and Gas Commission (BCOGC) extend their jurisdiction over pipelines entirely within their provinces.

Pipelines crossing provincial or international border are under the NEB's jurisdiction.

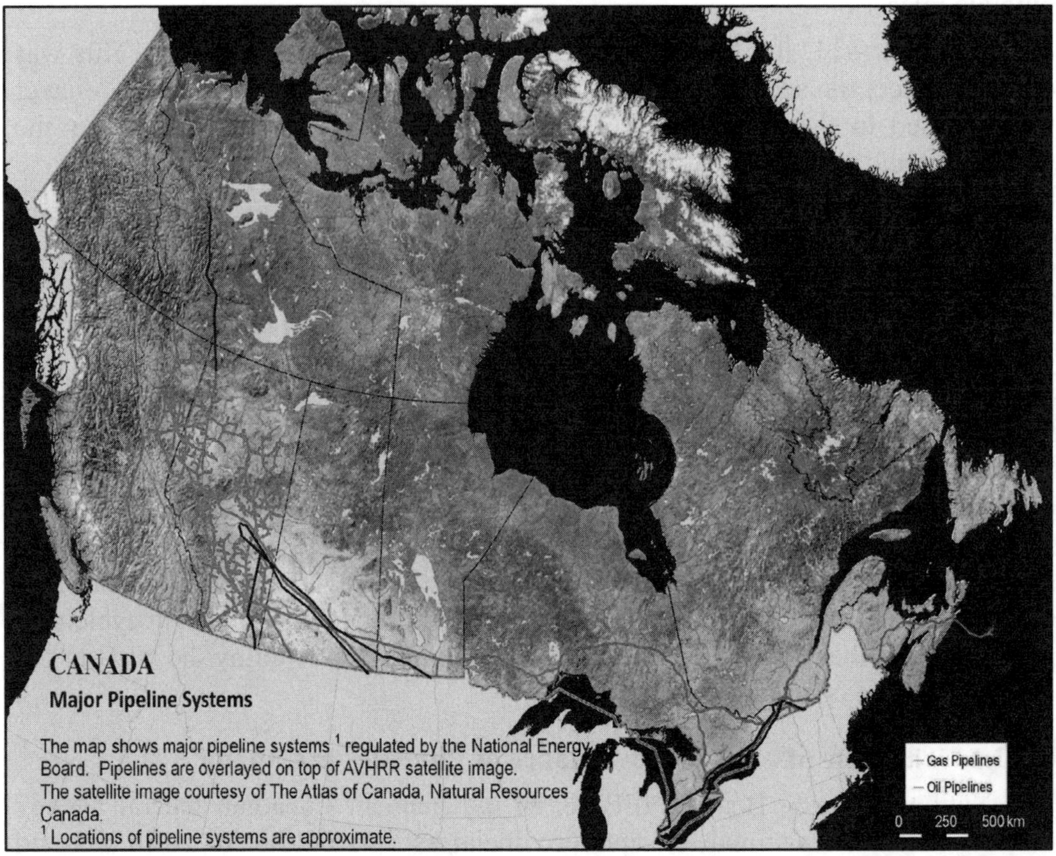

Figure 2: Major Pipeline Systems regulated by the NEB. Satellite image courtesy of the Atlas of Canada, Natural Resources Canada

The mandate of the NEB is wide and is guided by the basic principles of sustainable development. When making a decision all environmental, social and economic considerations are taken into account. This is reflected in the NEB's goals:

- Goal 1 – NEB-regulated facilities and activities are safe and secure and are perceived to be so.
- Goal 2 – NEB-regulated facilities are built and operated in a manner that protects the environment and respects the rights of those affected.
- Goal 3 – Canadian benefit from efficient energy infrastructure and markets.
- Goal 4 – The NEB fulfills its mandate with the benefit of effective public

engagement

- Goal 5 – The NEB delivers quality outcomes through innovative leadership and effective support processes.

GIS, in particular, supports Goal 1 and Goal 2 as described in the following sections.

6.2.2.3 The Evolution of GIS Program

Implementation of a GIS program within an organization often results in much higher costs than originally anticipated. The costs are of two major kinds; an initial cost to set up a program, followed by an operational cost, often significant, to maintain data and software in addition to staff or contractors. Recognizing this, the NEB was not an early adopter of GIS. The first step towards an enterprise GIS was a feasibility study conducted in 2003. The study identified potential use cases of GIS within most of the NEB's business units and set a roadmap for implementation. For a program to be successful there has to be a strong program sponsor, and at least one subject matter expert to turn that program into reality. The NEB had a strong program sponsor in that the leader of the Operations Business Unit had learned the value of GIS from his industry experience. A GIS Analyst was hired into a newly created position. The NEB also had a well designed system development methodology, experienced project managers, and an involved project working group. The NEB relied on the Geomatics Industry through contracts to fill other needs.

6.2.2.4 iMap – A web-based GIS

The initial focus was development of an intranet system, iMap, to show the NEB's regulated pipelines as illustrated in Figure 3.

The technology part of the GIS implementation turned out to be the easiest and within less than a year, iMap was up and running. Attribute data and/or mapping the NEB regulated pipelines, some of which were constructed even before the Board came into existence in 1959, has presented a real challenge. As opposed to other regulatory agencies, the NEB initially decided not to burden the regulated companies with requests for GIS data. Instead it tackled it in house by various means. Since many of the regulated pipeline systems are several thousand kilometers long, the NEB approached the Canadian Energy Pipeline Association (CEPA) to gain support in having companies operating large pipeline systems to provide GIS data on a voluntary basis. To date through the NEB internal initiatives and the help of CEPA the NEB has most of its regulated pipelines represented in iMap.

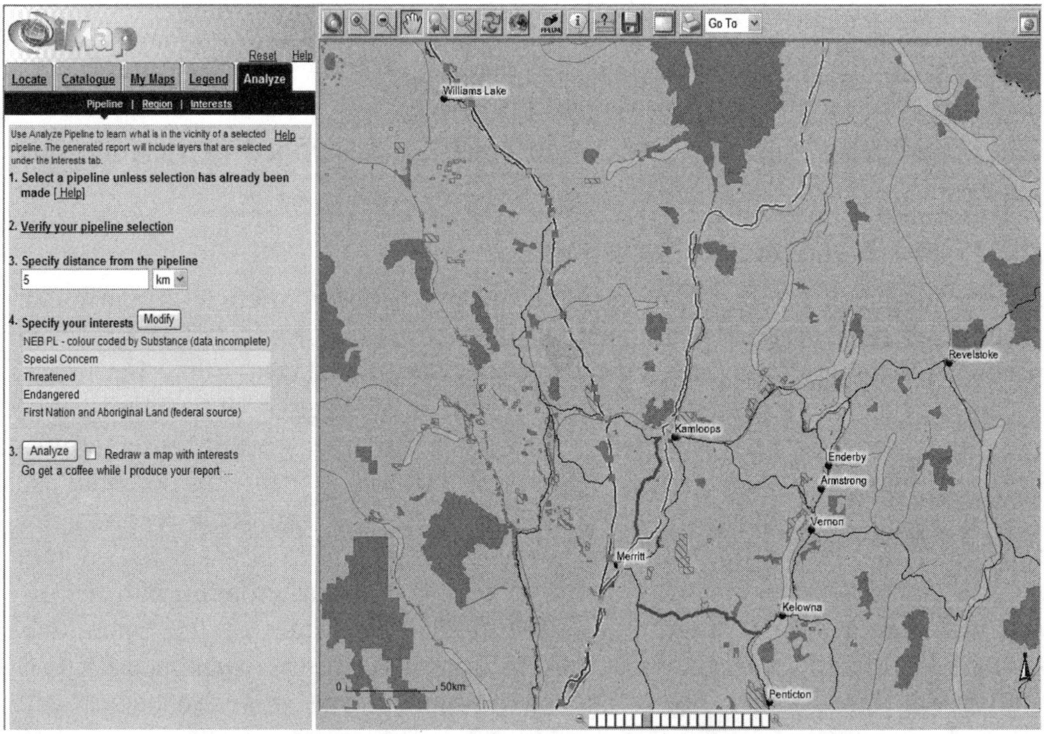

Figure 3: iMap, a web-based system showing NEB regulated pipelines in relation to populated places and parks. (Figure courtesy of the NEB)

With iMap the NEB staff could locate a pipeline, visually examine its vicinity, and conduct simple geospatial analyses. For example, iMap can generate a list of populated places, First Nation reserves, or other pipelines within a certain distance of a pipeline. GIS analytical capability was further leveraged to support a Risk-Based Lifecycle approach to regulations through development of consequence models. These will be discussed in further detail in section 6.2.2.6.

6.2.2.5 GIS and integration with other Information Systems

It is important to recognize that GIS provides only one aspect of data; geospatial location. As much as geospatial location is important, core information tends to exist in other information systems. This is also the case at the NEB. To cite just a few examples, information on incidents, results from inspections and audits, list of company emergency manuals, facility data, Aboriginal profiles are all collected in independent information systems or databases.

iMap, the GIS web-based system, was designed to be easily integrated with other NEB systems through HTML protocol. This means that iMap can accesses other information systems, and other systems can access iMap as shown in Figure 4.

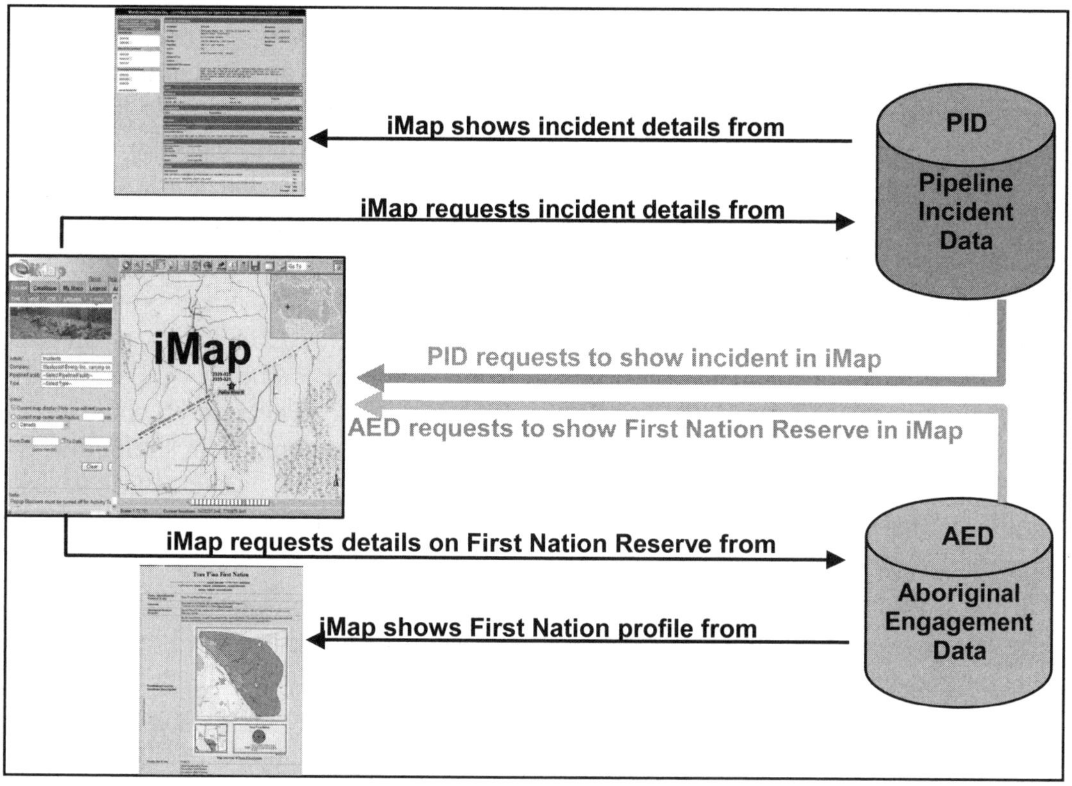

Figure 4: iMap accessing data from PID and AE, and vice versa (Figure courtesy of the NEB)

For example, when examining information on incidents in the incident database, a user can display the location of this incident in iMap. Conversely, a user can click on an incident in iMap and read all the details about the incident that are stored in the incident database. Integration of GIS with other information systems has been implemented in a phased approach, starting with geo-referencing of legacy data first, and only then proceeding to the actual system integration. As of 2009, systems containing information on incidents, encroachments, aboriginal communities, and emergency manuals have been integrated. When required, the NEB also leverages available high resolution satellite images provided by Google, as obtaining and managing satellite images for all of Canada is cost prohibitive.

The flexible communication of GIS with other information systems has proved to be the most useful aspect of the overall system.

6.2.2.6 Risk-Based Lifecycle

With the introduction of Onshore Pipeline Regulations in 1999, the NEB has moved from prescriptive to goal-oriented regulations. Prescriptive regulations are essentially a check-box of what's and how's. They specify means by which compliance with legal regulatory requirements is achieved. In contrast, goal-oriented regulations set clear outcomes, but allow companies to decide how best to achieve these outcomes. This approach allows each company to develop innovative and cost-effective management system programs that make the most sense in their particular circumstances. The Board then focuses on the overall effectiveness of the programs designed by the regulated companies.

To further improve and innovate the way the Board does business, since 2003 significant efforts went into development of the **Risk-based Lifecycle** approach to regulatory oversight. This is a process by which the NEB can direct its resources towards companies posing the greatest regulatory risks. The **risk** is considered during the entire **lifecycle of a pipeline**, from the moment such a pipeline is planned by the applicant to the day it is no longer needed. A traditional risk model is based on the formula, Risk=Probability x Consequence with risk represented by a single number.

For the NEB, the **probability** is measured by the adequacy, implementation and effectiveness of the company's programs. **Consequence** is the result if an incident occurs, and what are the impacts and severity of impacts. Would the public or employees be affected or concerned? What would be the potential impacts on the environment? A company's risk is plotted on a risk matrix shown in Figure 5 that is segmented into high, medium and low depending on the probability and consequence scores. The matrix helps to determine the types of NEB compliance activities, such as inspections or audits, depending on where the company's risk is in the matrix. The high risk zone will result in more compliance activities.

GIS contributes to Risk-based Lifecycle by calculating consequence. Four GIS consequence models where developed, each for different product transported via pipelines. Products such as natural gas, sour natural gas, natural gas liquids, and liquids, interact differently with the environment. For example, an oil release in the proximity of water is significant, because an oil leak will affect the watershed. A gas release is of insignificant consequence to the watershed as it will evaporate into the air. Proximity to population is significant for both products, although more so for gas, because of the potential explosion and heat radiation. Sour gas release has even more impact on population due to its poisonous nature.

The GIS consequence models utilize land use, significant areas like parks, permafrost, hydrographic network and population density layers, to segment all the NEB regulated pipelines. Each pipeline segment is assigned a consequence class. Consequence classes for each segment are further combined with company performance data such as number of incidents, effectiveness of management system, etc. to form companies' relative risk assessment.

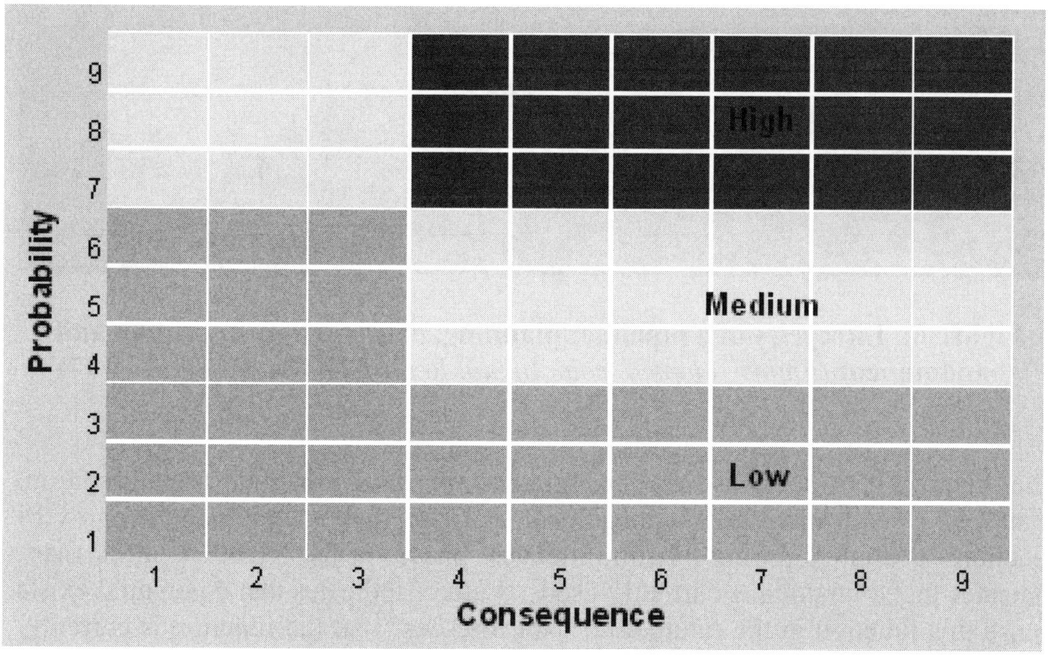

Figure 5: The NEB risk matrix, with high, medium and low zones. (Figure courtesy of the NEB).

GIS in itself raises additional challenges. Positional accuracy of the data used for consequence analyses varies. The NEB is continually improving the accuracy of pipelines, but still does not have a complete pipeline layer. Other layers used in the consequence analyses, such as land use, hydrographic network, etc, are maintained by other agencies'. Completeness, currency, and accuracy of these layers vary. Currency is a particular problem when it comes to the land use and permafrost layers. Another impact on the models is whether the data is fit for use. For example, the best population density dataset for Canada is available from Statistics Canada. Population is reported within a census block. However, census blocks vary in size, representing a city block in the cities, and much larger area outside of cities where population density is low. The consequence model makes an assumption that population density is uniform within a block which may or may not be true.

When interpreting risk assessment it is very important to ensure that users understand the limitations of the model and data, and that the output from the model is only as good as the data entered into it.

6.2.2.7 GIS and the Pipeline Lifecycle

At the NEB GIS is used within the entire lifecycle of a pipeline, as shown in Figure 6.

Figure 6: Lifecycles of a pipeline: planning, construction, operation and abandonment *(Figure adapted from Annual Report 2008 to Parliament)* **(7)**

The major GIS functions can be grouped in data creation, transformation, analyses, visualization and management. Table 1 below shows major GIS functions as applied during pipeline lifecycle phases at the NEB. A check mark indicates that a function is currently used. 'Future' indicates that a potential exists to use this function in the future, and 'Not foreseen' that the function is currently not foreseen for the future use.

Table 1: Major GIS functions applied to pipeline lifecycle phases at the NEB

Pipeline Lifecycle phase	GIS Data Creation/ Update	GIS Data Transformation	Analyses	Visualization	GIS Data Management
Planning	✓	✓	✓	✓	✓
Construction	Future	Future	Future	✓	Future
Operation	Future	Future	✓	✓	Future
Abandonment	✓	Not foreseen	Not foreseen	✓	Not foreseen

During the **planning phase** all GIS functions are used. When a company applies to the NEB to construct new facilities, modify existing ones, or purchase or sell asset, GIS data is updated accordingly. New facilities are digitized from the application provided by the company and attribute data are updated. In some cases, the application approval process is supported by geospatial analyses. A simple example involves locating a facility in question and examining its proximity for closest populated places, aboriginal lands, species at risk, other pipelines in the vicinity, etc. More involved applications lend themselves to more complex analyses. In 2008 the NEB received an application for the 149.7 km long sour gas Redwillow Pipeline by SemCAMS Redwillow ULC. GIS was used to assess potential impacts in the pipeline vicinity in case of an incident. GIS also contributed to understanding where other sour gas pipelines currently exist and who their operators are. Finally, GIS facilitates effective communication. Creating maps or videos, showing flyovers along proposed pipelines greatly helps project working groups and the Board Members to understand the geographical context of the application.

Once the application is approved and the **construction phase** begins, GIS is used to efficiently plan inspections performed by NEB personnel. For example, in 2008 three major projects reached the construction phase in Canada: 1,078 km Alberta Clipper by Enbridge, 288 km Southern Lights by Enbridge, and 1,235 km Keystone Pipeline by TransCanada. Having all three projects on a single map helped to develop an efficient inspection schedule. The NEB has also experimented with geo-tagging inspection photos. The photos can be incorporated into interactive report and geotagging gives the ability to easily return to the place originally visited.

At present, the largest usage of GIS is seen during the **operation phase** of the facilities. It includes locating of incidents, landowner complaints, or encroachment of third parties on pipeline right-of-ways. When the locations are known, GIS helps assessing potential consequences. In addition, the NEB inspectors use iMap's analytical functions to understand geographical contexts around pipelines while preparing for inspections, or when working on companies' audits. Finally, as described earlier, GIS is used to develop consequence rating of facilities, used in the company risk prioritization model.

Lastly, when facilities are no longer needed and the company proceeds to the **deactivation, decommissioning** or **abandonment phase,** the NEB needs to understand where these particular facilities are located. In the recent years the NEB has seen more interest from the public about locations of abandoned facilities.

6.2.2.8 Case Study: Alberta Energy Resources Conservation Board

We have chosen the Canadian province of Alberta for a short case study for several reasons:

- Alberta is one of Canada's most energy-rich provinces, with a large percentage of the nation's pipelines originating in, or traveling though the province.
- It is also a jurisdiction with some of the greatest concentration of pipelines in all the world.
- The Alberta Energy Resources Conservation Board (ERCB) has a strong GIS-based data management system and a history of innovation in this field

The Energy Resources Conservation Board (ERCB; www.ercb.ca) is an independent, quasi-judicial agency of the Government of Alberta. The Mission of the ERCB is to ensure that the discovery, development and delivery of Alberta's energy resources take place in a manner that is fair, responsible and in the public interest.

The ERCB regulates the safe, responsible, and efficient development of oil, natural gas, oil sands, and coal, as well as the pipelines to move the resources to market. Regulation is done through two core functions: adjudication and regulation, and information and knowledge. ERCB approval must be given at almost every step of an energy project's life.

The ERCB regulates (2006 figures):

- 159,500 operating natural gas and oil wells
- 33,700 oil and gas batteries, plants and other facilities
- 392,000 km of pipelines
- 12 producing coal mines
- 38 commercial oil sands plants

The ERCB is responsible for managing Alberta's energy resources, including:

- 174 billion barrels of bitumen
- 1.6 billion barrels of conventional oil
- 40 trillion cubic feet of natural gas
- 34 billion tons of coal
- 877 billion cubic feet of recoverable coal bed methane reserves

The government of Alberta owns about 80 per cent of the province's mineral rights, such as oil, natural gas, coal, and the oil sands. In other words, most resources are owned by the people of Alberta through their government. While private companies can develop these resources, the ERCB is authorized by the government to protect the public's interest relating to the discovery, development, and delivery of these resources. Regulation is needed so that non-renewable resources are produced in a safe, responsible, and efficient manner, without waste. The ERCB also ensures that everyone affected by development has a chance to be heard. When conflicts regarding development remain unresolved between companies and landowners, the ERCB works to settle the issues in a fair and balanced manner.

The ERCB has long been a user of computer assisted mapping technology, particularly with respect to mapping the location of pipelines. GIS was introduced as a technology within the ERCB in 2002. Initially GIS technology was used to replace a mainframe pipeline mapping system that had become difficult to support. Since the early introduction of GIS technology at the ERCB, GIS has evolved into an enterprise-wide program supporting the day to day business functions of many areas of the Board. GIS is becoming a core technology and a target technology at the ERCB. GIS is currently used in a number of business processes, systems, data, and other software applications. It is becoming an integral part of the computing infrastructure contributing to the success of the ERCB's service delivery and operations. While close to half of ERCB staff have access to GIS technology at the desktop, the direction for future solution development will be the web-based environment. To that end, the ERCB has developed a web-based spatial data viewer that can be easily configured to meet the specific requirements of any business unit.

The ERCB provides a streamlined approach to processing some 30,000 energy development applications each year. Staff handle project reviews, audits, and approvals related to new or modified oil and gas facilities, such as wells, pipelines, batteries, and gas plants. Most applications are submitted by operators through an electronic application submission process. The current pipeline application submission processes remains heavily paper-based, although efforts are currently underway to upgrade this submission process as well. Operators planning a pipeline submission must obtain a plat of the area of interest from the ERCB; this plat is generated from the ERCB's enterprise GIS environment. The operator will indicate on the plat the proposed route of the planned pipeline and submit the plat as part of the application. Once the application has been received by the ERCB, reviewed by the appropriate staff and other government departments, and approved, the proposed route of the pipeline is entered into the enterprise GIS environment's database.

Directive 56 (which stipulates the information industry must submit as part of the pipeline application process) states:

> The ERCB uses pipeline base maps to show the approximate location of licensed pipelines under its jurisdiction. The maps are produced on an individual township basis; if a township is very congested, a larger scale is used, e.g., quarter section. The base plan maps show the best pipeline location data available to the ERCB and identify the licence number of each pipeline. Under no circumstances should these maps be used for physically locating pipelines, as they only provide a general representation of the pipeline route. When an application is approved, the ERCB transfers the base map information to the master graphics file, which is available to all users. The base maps provided by ERCB Information Services are used to digitize the location of pipelines and must be at an 11 x 17 inch format. The ERCB does not require that the applicant submit as-built or survey drawings, but does require that the accuracy of plotting pipelines on the base map reflect the actual surveyed route.

While the responsibility for conducting safe, efficient operations rests with each operator, the ERCB carries out investigations and inspections to ensure that this is done. The ERCB maintains staff in Field Centres that conduct thousands of inspections of wells, facilities, and pipelines throughout the province each year.

Field staff inspect construction, operation, and abandonment operations at oil, gas, and oil sands facilities, respond to emergencies and public complaints on a 24-hour basis, facilitate resolution of landowner-industry conflicts, participate in public-industry liaison committees, and ensure a consistent approach to noncompliant operators. One of the main tools used by inspectors for these tasks is a GIS-based field inspection system. Graphical visualization integrated with the FIS Corporate Database greatly benefits the ERCB Field Inspector's ability to perform their tasks. Viewing a geographical area and associated surveillance information provides understanding, resulting in improved accuracy and integrity of inspections conducted. The system permits the inspectors to accurately locate facilities to be inspected, and demonstrate the positional information of facilities to operators and the public. Additional functionality within the inspection system provides the inspector with the ability to collect positional information using GPS.

6.2.3 Other Countries

6.2.3.1 Australian Energy Regulator

The Australian Energy Regulator (AER, www.aer.gov.au) is a constituent part of the Australian Competition and Consumer Commission (ACCC). It was established under Part IIIAA of the Trade Practices Act 1974 and operates as a separate legal entity. The AER is responsible for the economic regulation of gas transmission and distribution networks and enforcing the national gas law and national gas rules in all jurisdictions except Western Australia.

The AER's functions and powers in relation to the economic regulation of covered gas transmission and distribution pipelines include:

- approval of certain access arrangements required to be submitted by service providers under the National Gas Law and National Gas Rules
- review of annual reference tariff variations in accordance with relevant access arrangements
- annual monitoring of compliance of service providers' obligations under the National Gas Law and National Gas Rules
- undertaking enforcement functions as required in relation to breaches of the National Gas Law, National Gas Rules and Regulations
- hearing disputes in relation to the terms and conditions of access for relevant pipelines
- approval of competitive tendering processes and terms and conditions of access for competitive tender pipelines as required under the National Gas Law and National Gas Rules
- other functions and powers required to be undertaken under the National Gas Law and National Gas Rules including those associated with the Gas

Market Bulletin Board

Pipeline regulation within Australia is often the responsibility of various state agencies. In the Department of Employment, Economic Development and Innovation (formerly the Department of Mines and Energy) for the Queensland government (www.dme.qld.gov.au), the Senior Spatial Information Officer with Mapping and Survey Services enters spatial pipeline data into the corporate graphical database "MERLIN".

Pipeline applications are received in the department's Petroleum and Gas tenures area. Digital data is submitted with each pipeline application, normally in ESRI Shapefile format. The digital data indicates the location of the approximate centerline of the proposed pipeline. This digital data is checked in ArcGIS and then imported into the MERLIN database. Once the pipeline is constructed, there is usually an easement survey of the pipeline sent to the department, and the MERLIN database is updated to adjust the location of the pipeline to this surveyed information.

The timeline for this process could be years and the difference between the original pipeline position and the constructed position can be substantial particularly in years gone by. The department strives to keep their data as accurate as possible by insisting on accurate digital data from the applicant in the first place. Many of the applications received are now based on GPS positioning.

6.2.3.2 United Kingdom and Ireland

In Great Britain, the Health and Safety Executive (HSE, www.hse.gov.uk) is responsible for the health and safety of the citizens of Great Britain. Part of that responsibility involves the regulation of pipelines within the country and territorial waters. HSE believes that where pipelines are properly designed, built, operated, maintained and eventually decommissioned they are safe for the pipeline workforce and the general public. The regulation of pipelines in Great Britain is covered by the Pipeline Safety Regulations 1996 (PSR) and of particular concern are those pipelines known as major accident hazard (MAH) pipelines. The PSR applies to pipelines in Great Britain and to those in territorial waters and the UK Continental Shelf.

Major accident hazard pipelines (MAH pipelines or MAHPs) include the high pressure natural gas supply transmission and distribution network. They also cover other pipeline systems transporting oils, chemicals and other gases. Examples include ammonia and ethylene. The responsibility for enforcing health and safety law for onshore and offshore MAH pipelines forms part of the role of the Specialized Industries Gas and Pipelines Unit in HSE's Hazardous Industries Directorate (HID, www.hse.gov.uk/hid). The unit and its pipeline inspectors achieve regulation and compliance through the enforcement of the Health and Safety at Work Act 1974.

The Pipeline Safety Regulations 1996 (PSR) place a duty on operators of MAH pipelines to provide certain information to HSE at various stages in the lifecycle of a MAH pipeline. Particulars to be included in the notification relating to construction of a major accident hazard pipeline include the proposed route of the pipeline in the form of maps or drawings. The notification of a change to the route or position of a pipeline also requires the particulars in the form of maps or drawings of the new route or position.

National Grid (www.nationalgrid.com/uk) is an international electricity and gas company. National Grid owns and operates the National Transmission System throughout Great Britain and owns and operates a significant Gas Distribution network throughout the heart of England. National Grid is responsible for the safety and maintenance of the 18 000 km long high-pressure transmission network that supplies gas to business and domestic consumers across the UK. Health and Safety Legislation in the UK requires National Grid to demonstrate that the risks to individuals living and working in the vicinity of gas pipelines are adequately managed. In order to enhance its capabilities for assessing the risks due to geological ground conditions, National Grid required detailed information about the distribution and potential severity of geohazards across the UK. The British Geological Survey (BGS), as the primary holder of national geological hazard data in the UK, were commissioned to design suitable GIS layers as part of the risk assessment methodology. BGS and National Grid have produced a new GIS for the rapid assessment of the potential for a significant geohazard to adversely affect any section of the UK natural gas pipeline transmission network.

Bord Gais Eireann (BGE, www.bge.com) is a gas utility responsible for the transmission and distribution of natural gas in Ireland. BGE had successfully implemented CAD into their design offices and began looking closely at the state of GIS in the early 1990's. But it was not until the late '90s, after extensive research, that BGE first introduced GIS to create an accurate representation of the gas network from a large collection of paper maps and documents. The integration with Ordnance Survey of Ireland spatial data and GeoDirectory digital data enhanced processes and continues to provide higher quality service to a growing customer base in an efficient, safe and cost effective manner through rapid access to up-to-date network information across the organization. In BGE today, geo-spatial technology is reaching far beyond traditional GIS. By integrating disparate datasets, GIS provides visualization and analysis that illustrates relationships, connections and patterns that are not obvious until displayed graphically. This enables BGE engineers and managers to make more informed decisions.

Legacy systems, including the IUS (Integrated Utility System), MRAM (Mains Risk Assessment Methodology) databases have been integrated with GIS to provide job management details and hazard analysis to assist network Renewals. Engineers can visualize the expected performance of the gas network under any given condition by importing pressure results from the network analysis system to GIS.

Responding to 'Dial before you Dig' requests from external parties has become very efficient and enhanced safety. Issuing network information in digital format to large utility groups substantially reduces the number of queries. Field personnel now have instant access to network information on laptops. An easy to use application provide simple red lining facilities to external personnel involved in the utilization of the gas network – such as maintenance crews, sales personnel, etc.. Cost effective access to data has been provided by GIS Web on the BGE Intranet. BGE staff now has the facility to view other utilities network data in their GIS.

Having delivered all anticipated benefits including a significant reduction in the number of designers, the GIS strategy was modified to focus on applications to increase support for business processes such as asset maintenance, operational services, planning and engineering, sales and marketing. These solutions were adopted by other BGE companies such as UGI, Firmus, Aurora and Networks Transmission with future applications planned for Networks Transportation.

Applications, never before considered, have been made possible. Digital video, recorded on helicopter patrols, is integrated with pipeline and background map data to allow safety engineers to identify, review and record areas of potential concern. BGE can now visually analyze customer penetration, identifying potential customers adjacent to current networks. Indeed, the potential exists to enable a spatial view of any customer related data in GIS; large customers, vulnerable customers, or customers managed by different shippers.

6.3 Future Trends

Pipeline regulatory agencies around the world are at different states in the use of GIS technology within their agencies, and at the requirement for the submission of GIS-ready spatial data as part of the pipeline application process or compliance monitoring and field inspection processes. But there is no doubt that there is much larger potential for the use of GIS and other related spatial technologies in the pipeline regulatory process. Future trends include – but are certainly not limited to – the following.

- Not all pipeline regulatory agencies world-wide currently have an electronic submission process for pipeline applications. A future trend for some regulatory agencies would include the development of an electronic submission process that would include not only pipeline spatial data in GIS-ready format, but appropriate spatial metadata as well.
- Following from the previous point, regulatory agencies would benefit from processes that could automate – to the extent possible – the review and approval of the application based on a review of the spatial data provided by the applicant. A major component of the approval process would include the automatic update of the regulatory agency's GIS

database with the information submitted as part of the application process; the integration of the submitted data with the existing pipelines using extensive quality control processes to ensure appropriate connectivity, etc.

- A major component of – and a major expense of – any GIS-based solution is the acquisition of, and use of, high quality, positionally accurate spatial data. Many regulatory agencies have created or acquired pipeline spatial data from historical records whose location accuracy may be suspect. Updating the location accuracy of pipeline spatial data will be time consuming and costly. The use of right-of-way spatial data and/or GPS technology to correct pipeline locations may require changes to the submission regulations of some regulatory agencies; regulatory agencies may consider requiring the submission of as-built spatial information after pipeline constructions is completed and when operational activities have necessitated changes to the location of portions of a pipeline.
- There is great potential to use of GIS technology and multiple – potentially new – data sources to ease/facilitate the application approval processes and the compliance monitoring processes. By integrating disparate data sets, GIS provides visualization and analysis that illustrate relationships, connections and patterns that are not obvious until displayed graphically. The use of current imagery data can enhance the view of relationships between the pipeline and surrounding environment.
- GIS is more than just electronic mapping, and can be used for more than just facilities management. GIS technologies and spatial data should be pervasive through all business areas of pipeline regulatory agencies. The GIS data store should become the single data source for all regulatory processes, and incorporated in the analyses and decision-making processes of the agency.
- The use of GIS technology within regulatory agencies often focuses on the current state of the pipeline network. Many regulatory agencies could benefit from including the temporal aspect of pipeline data in their GIS data structures. The ability to track a pipeline's history of operators, commodity carried, inspection records, and problems over the years will enhance a regulatory agency's ability to form policy which targets the areas of greatest concern.
- Many regulatory agencies could benefit from an internal portal providing access to pipeline spatial data as well as information related to pipeline operators and operations through a geospatial interface;
- Similarly, the public in the jurisdictions regulated by pipeline regulatory agencies are more aware of the impacts of pipelines to the environment and the public in general. They are demanding an active role in the

regulatory process. Many regulatory agencies are actively seeking public participation in hearings. A public portal displaying general locations of the active applications, construction projects, and operating pipelines would increase public awareness.

- Geospatial indexing or geo-tagging of public records, pictures and other documents to specific locations along the spatial representation of a pipeline would provide staff of regulatory agencies access to additional information for their analysis and decision making processes. Providing one location – one system – whereby the necessary information could be viewed and accessed, will improve the performance of agency staff by reducing the number of data sources to be reviewed during a decision making process.
- Many agencies use field inspections as part of the regulatory process. Not all use GIS-based field inspection systems. Given the spatial nature of pipeline data, it is a logical step to distribute pipeline spatial and attributed data to the field. The collection of additional spatial data during field trips or the field inspection process would aid in improving the data content and quality of the agency's pipeline spatial data store. Data collected in the field should be synchronized and uploaded to the regulatory agency's corporate spatial data repository.

Technology of the 21st century offers all the building blocks to address potential uses. However, technology is only one contributing factor. Other important factors are; continuous process improvement, a continuing investment in data, on-going information management, and system integration.

References

(1) Yoon, M. S., Warren, C. B., and Adam, S., "Pipeline System Automation and Control," Chapter 8, ASME, New York, 2007

(2) Johnson, D., "The Rules Covering Public Awareness and GIS: A Pipeline Safety/Compliance Perspective" Proceedings of GITA's 15th Annual GIS for Oil & Gas Conference and Exhibition. Calgary, Canada, 2006

(3) Veenstra, P., "Meeting Future Challenges of Pipeline Data Management" Proceedings of GITA's 15th Annual GIS for Oil & Gas Conference and Exhibition. Calgary, Canada, 2006

(4) Hall, S. (2003) The National Pipeline Mapping System – A Review. Business Briefing Exploration & Production: The Oil & Gas Review, Vol. 2, pp. 84.

(5) In this chapter, pipeline refers to a pipeline and all related facilities such as compressor stations, meters stations, valves, etc. Facility is another term used to refer to pipelines.

(6) The value of shipped products is derived from annual throughputs and the

weighted average yearly commodity price. This value reflects shipping over 45 000 km of pipelines in 2008, rather than 70,000 km. In 2009 a large 25 000 km system of Nova Gas Transmission Limited in Alberta was transferred from the ERCB to the NEB jurisdiction. (Data from NEB's 2008 Annual Report.)

(7) National Energy Board, 2009. Annual Report 2008 to Parliament

Acknowledgments

The authors of this chapter gratefully acknowledge the contributions Brendan O'Shaughnessy, GIS Manager, Bord Gais Eireann to portions of this chapter.

7. Geomatics Implementation

In order to round-out the discussion on geomatics, we feel that it is helpful to include some guidance on implementing geomatics into projects or an enterprise. Introducing new views, processes, or technology in general is always uncomfortable. Our intention with this chapter is to provide a framework and general checklist of some of the strategies and pitfalls when it comes to implementing geomatic technology and processes. Hopefully, this way geomatics does not seem so distant and mysterious to new users and collaborators.

In our discussion, we consider an implementation as a move towards spatial data management and the geomatic tools and processes that would support engineering. Simply using GIS software to make maps would not be considered an implementation. The following discussion is broken into two main parts; first, organizational or enterprise implementations and second, implementation for pipeline projects.

7.1 Organizational and Large Implementations

The breadth and diversity of situations that could be encountered during large implementations are considerable. In fact, there is so much variety from one situation to the next and from organization to organization that an entire book could be devoted to this one topic alone. Nonetheless, there are some fundamental and common themes that all implementations will involve. Rather than provide a wide-ranging list of all potential scenarios, our goal in this section is to enlighten the reader on the fundamental concepts, provide a general framework, and a checklist at the end.

Implementing geomatics has for some, proven to be a challenging task. For others harnessing the benefits of this discipline into existing workflows and processes has been a much smoother and satisfying process. The difference between the two outcomes can, in most cases be attributed to a lack of patience and proper planning. It can be an extensive and in some cases ongoing task. As has been discussed in the first chapter, Geomatic concepts can be applied within the entire pipeline value chain. For this reason it is very easy to get carried away with trying to implement too many things at one time. As with most new innovations, the key to a successful geomatics implementation is a well thought out plan and a well managed execution of that plan.

Further insights are gained from diligently capturing processes, problems, solutions and successes along the way so that pitfalls and triumphs experienced the first time around can either be avoided or targeted the next time. Although this book is dedicated to geomatics within the pipeline sector, most of the concepts discussed in this chapter are generic enough that they are quite suitable for any kind of a geomatics implementation. This was done mainly to demonstrate that these concepts are common to all geomatics implementations and in doing so provide some perspective on the global capabilities of geomatics both within and beyond the business of pipelines.

At the most basic level, the implementation of geomatics is about people, technology and data (Figure 1). Managing these three elements successfully is the key to making or breaking an implementation. The right combination of these three building blocks will ensure a solid foundation for a successful geomatics platform.

With each one of these elements, there are key fundamentals that require the precise amount of attention and consideration. For anyone familiar with project management or technology implementations, these should come as no surprise. None the less, too little attention is often devoted to at least one or more of these areas. This creates a great deal of conflict and in the worst cases, can contribute to the complete failure of a project.

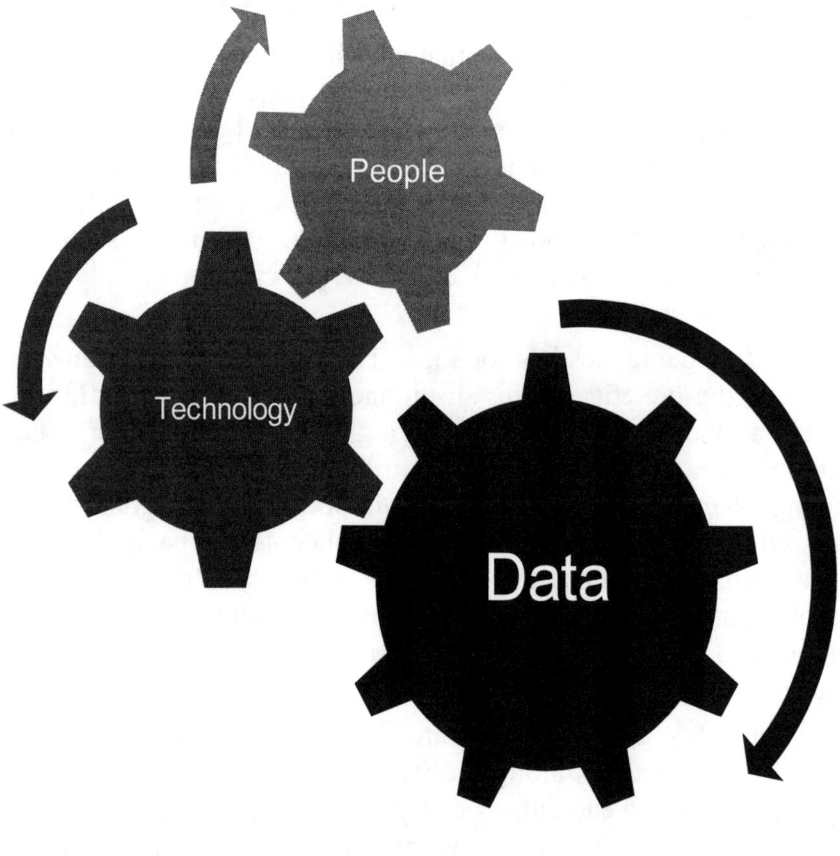

Figure 1. Conceptual framework of the major parts of a geomatics implementation

7.1.1 People

In many ways the most important part of any geomatics project is the people involved. Without the right people in place, the project has little hope of either getting started or being successful going forward.

This of course begs the question; *who are the right people?* In the case of introducing geomatic concepts to either a project or to an organization, there are key individuals that must first be identified. Likely the most significant people are what we call *project sponsors*.

Project sponsors are like minded individuals from a variety of different areas and levels that will support your geomatics implementation within the organization. Some you will already be aware of, others you will need to find, and still others you will need to recruit. In any event, these individuals will become the frontline and behind the scenes support network that will give you and your project credibility within the organization. Before anything else happens, getting your support network in place should be a first priority.

As mentioned, sponsors will come from different areas and different levels within an organization. This is an important point to recognize. Having your immediate co-workers as your entire team of sponsors is great but will do little for your project outside of your immediate area within the organization. To be truly effective you should aim to have sponsors from as many different business areas and levels of the organization as possible.

Having sponsors from different business units or other areas of the organization will impress upon senior decision makers that your proposed geomatic implementation will provide benefit across a wide range of groups within the organization. Furthermore, allies within different area of the organization are invaluable when the time comes to gain access to information and disparate datasets.

Vertical sponsorship within the organization will facilitate faster decisions and a much easier forward progression of the project. With at least one member of the senior executive team in support of your initiative, everything from getting necessary funding to securing the right audience for sales and idea pitches will be much easier.

By this point, it should be fairly clear *who* the right people are at the outset of a project or organizational implementation. Equally important, are the key individuals that need to be in place during the implementation once all of the approvals have been given to go ahead.

Making sure that the right people are steering the project on the front lines is another consideration. The best candidates for this task are the "B Players" within an organization (1). These are your trusted employees that have a habit of always getting things done, often without any fanfare and recognition. Having people like this in charge of the actual legwork required for an implementation will inevitably increase the odds of a successful implantation.

Ongoing resources are probably one of the final *who* questions that need to be answered. Some of the people that will support the geomatics initiative once it is up and running will have already been identified. Others will need to be recruited. Either way, the continued success of the initiative will depend to a large extent on the people that remain in place once the implemented geomatics workflow, technology etc. is up and running.

As is evident, the *Who* question is an important one when it comes to identifying key people involved in a geomatics implementation. The next section will discuss some of the considerations required around the technology itself.

7.1.2 Technology

When it comes to making decisions about what technology is most appropriate, there are a few important considerations to make. Of course at the end of the day, technology is never static which does provide for challenges. Choosing a technology today that will meet with the ever changing needs of both people and organizations can be a daunting task.

Adhering to a few simple rules when making decisions around technology can alleviate some of the challenges that typically present themselves. For example, technology decisions should be based whenever possible on the most up to date and current solutions available. As was mentioned, technology is in a constant state of change and therefore any new installations should be as modern as possible.

A rule of thumb is that technology, and in particular applications, have on average a 2-5 year lifespan before they need to be replaced or seriously upgraded to remain fully functional with other systems and applications that have changed or are new. What this means is that in order to get the most out your investment in technology, a vast amount of research should be conducted.

Key questions to seek full answers to are; *what is available*, *what is currently in use* and *what are the emerging trends*. Having answers to all of these questions before a decision for a particular path is made will help with choosing a path that will provide greater value over a longer period of time.

Interoperability with other systems is also a key consideration. Increasingly, technologies within organizations are converging. Geomatics in particular is a driver of this emergence.

Another general rule to follow is to choose technologies that are developed on either a common or open source platform. Doing so will help ease integration with existing infrastructure. This is important for a couple of reasons. First, often times system interoperability issues will only become apparent after a new technology is up and running. The time and costs associated with tracking down issues like this and correcting them can be substantial. In some cases the issues cannot be corrected. This immediately has a very negative effect on both the value of the new technology and also affects its lifespan.

Solutions based on common or open source platforms can facilitate a much quicker integration process with other systems and applications (2). This streamlines the implementation process and reduces project scope creep that can leave a very bad taste in the mouths of project sponsors, particularly those approving the spending.

A third rule of thumb for choosing technology would be to choose solutions that are as universally acceptable and as easy to use as possible. The more people within the organization that can use the solution the more valuable the solution is to the organization. Moreover, the easier the solution is to learn, the greater the numbers will be in terms of new adoptees.

Despite the uncertainty of technology, several things are certain. Needs change, technology advances, and people require more, in terms of flexibility, capability and interoperability. Keeping the above rules in mind when examining geomatics solutions will help identify where the balance can be found between both what is certain and what is uncertain.

7.1.3 Data Assets

Data are the third key elements. The importance of good information cannot be understated. The Geomatics system within an organization will be relied upon to not only ingest data but also provide information. Above all else, the value of geomatics technology is only as good as the information contained within the system. Whether it is new information that goes into a geomatics system or information taken from other systems, all of the information needs to be qualified and verified.

Before information is even entered into a system, the information several questions need to be answered. Questions like; *what information will be added to the system? Where will it come from? How will get from where it is now into this new system? Will it have to be created new? What has to be done to it to get into the geomatics system?* give an example of the many important considerations for information that is chosen to be added.

Creating a data model for the system is an effective tool for managing geomatics information and finding answers to the above questions. In terms of qualifying the information, the data model also provides a means to identify targeted parameters for accuracies and data integrity.

In the case of internally sourced data, it is usually easier to identify the quality of information as a full or at least partial lineage is often more readily available. Furthermore, the data is more likely to be in a state that easily integrates into a new system. However this is not necessarily the case, especially when dealing with something like legacy information or information sourced from a legacy system.

For data that has been sourced from an external location, even more care should be taken to both fully understand the lineage if available and also qualify the information prior to integrating it into the geomatics system. The process of modeling data before integrating it can serve as a rigorous quality check process.

Finally, once all of the questions around information integrity and lineage have been answered as accurately as possible, metadata should be created. Building proper metadata will help with capturing the value from the qualification process and provide a means of ensuring that the limitations of any information in the system are fully understood going forward.

All of this effort might seem excessive. However, the integrity of any geomatics process or system relies on the integrity of the information contained within. Moreover, the expense and resources required to implement these types of systems deserve and require that information be as good as possible. This ensures that the value obtained going forward exceeds the investment incurred.

To summarize, managing people, technology and information effectively can be a challenging task in the context of a geomatics implementation. Finding the correct balance between all three elements is paramount in influencing success. The above mentioned strategies are certainly not comprehensive but do provide a baseline approach based on practical experience and proven successes.

7.1.4 Other Considerations

It is evident from the above text that there are many considerations when undertaking to implement geomatic concepts at either the project or enterprise level. Given the variety of both projects and enterprises where geomatics is being incorporated, it is beyond the scope of this chapter to thoroughly cover all of the potential scenarios one may encounter. None the less, there are some commonalities regardless of the project or enterprise that we will discuss for the purpose of giving a cursory introduction to the geomatics implementation process.

7.1.4.1 Tradeoffs and Compromise

Although people, technology and information are the key elements of any geomatics implementations, there are also some other considerations that are worthy of mention. For those familiar with project management the project triangle (3) is a common concept. For those unfamiliar, the concept is fairly straightforward. The following diagram in Figure 2 represents the project triangle.

The three corners of the triangle represent the tradeoffs available with all projects. For example, if a project is a complete bargain, it is unlikely that it will work well or provide a quick solution. Conversely if a project or solution is to be great it is unlikely it will be cheap or quick to implement.

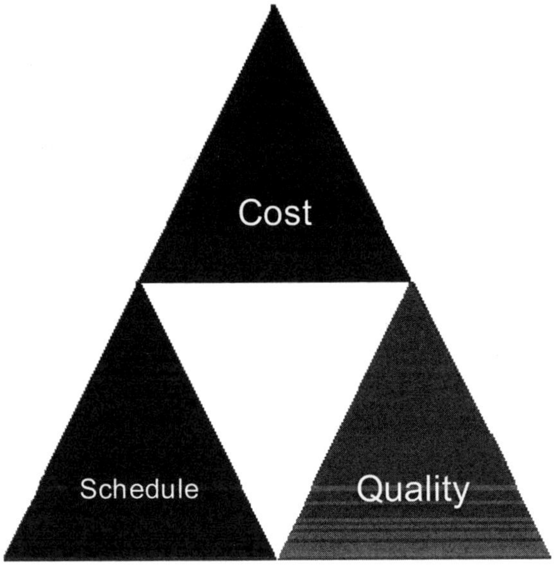

Figure 2: Project Triangle

All projects will fall somewhere within the triangle. Ideally the goal is to position your project exactly at the centre of the triangle. It is at this position where a perfect balance exists between having a solution that can be implemented quickly, is cost effective and works perfectly.

Unfortunately, most projects will be positioned between two of the three sides and not in the middle of the triangle. Thus solutions tend to be cheap and great but not quick, alternatively fast and great with the compromise being it is very costly.

Within the context of a geomatics implementation, one of the first decisions to be made is where your project will fit within the triangle. There is no wrong answer to this question. In fact, where a project or solution will fall within the triangle tends to be a function of the culture of an organization and its past experience with similar implementations. The key point to be made is an understanding that depending on where your implementation is positioned within the triangle, there are compromises that will have to be made. Having awareness of these compromises and being comfortable in this knowledge is really what is most important.

7.1.4.2 Managing Expectations

One final rule to remember is that it is very difficult if not impossible to please everyone. No matter how much effort is devoted to understanding needs and assessing goals, there will inevitably be people and things that are missed or pop up after the fact and have needs that are not satisfied.

The key to mitigating these unforeseen circumstances and dealing with this eventuality is to manage expectations and focus on positive benefits. Managing expectations is an important part of the pre-project and ongoing implementation marketing strategy. By being up front in the beginning and being open and honest throughout the implantation about the exact benefits and limitations of a particular solution, much less angst will be felt when unforeseen circumstances present themselves.

In short, if a solution is going to be cheap and fast to implement because of strict timelines and budgetary restrictions, it is critical that the compromises around how good it will be are openly discussed. Essentially, full disclosure and openness is the best way to manage expectations and ensure that what gets implemented is meets with the expectations of the organization.

7.1.5 Implementation Checklists

As a final tool, the following checklists will provide the reader with an insightful guide for some common considerations at all three levels of implementation; Pre-implementation, Implementation, and Post-implementation. Again, this is not meant to be a comprehensive list but an introductory guide that can be used as a starting point for understanding the fundamental steps involved in an implementation.

The following checklists are intended to provide the initial picture that can be taken and built upon to suit the varied needs of each individual reader. Planning the plan and the plan are the critical first steps in a geomatics implementation (4). These checklists provide an inside look and form the basis for a preliminary plan.

7.1.5.1 Pre–Implementation

As has already been discussed, the pre-planning steps are critical for ensuring a smooth and successful outcome during actual implantation. Proper planning up front helps alleviate implementation woes.

The following twelve steps provide a basic starting point for a pre-planning exercise. These steps are certainly not comprehensive but will provide the reader with a basic understanding of some of the more critical considerations and decisions that need to be made leading up to the implantation of a geomatics solution.

1) Needs Assessment – Identify clearly what you are trying to accomplish or what the problem you are trying to solve is.
2) Create a preliminary project charter that describes in details the problem and solution.
3) Business Case – Align your project with the larger organizational goals and strategies.
4) Decide whether the solution will be sourced from internal or external resources.

5) Create and distribute RFI (Request for Information) to potential solutions vendors.
6) Review RFI responses and adjust your needs assessment accordingly.
7) Decide on either internal or external solution.
8) Seek management approval to proceed.
9) Decide on whether or not a Pilot Project is warranted or possible.
10) Choose solution/solution provider.
11) Create the final project charter that describes in fine detail the project in its entirety.
12) Create Implementation schedule and proceed with project.

7.1.5.2 Implementation

During the actual implantation a number of common strategies will help with making sure that an implementation is as painless as possible. Following the steps and points outlined below will alleviate the potential for major project issues and conflicts.

1) If possible conduct a Pilot project at a smaller scale.
2) Review pilot and adjust main roll out strategy.
3) Implement the project in a Phased roll out.
4) Conduct frequent status updates and team meetings to share knowledge and keep everyone informed.
5) Communicate all project updates to all project team members.
6) Keep senior management informed of all progress, good or bad.
7) Change management – People element – stay in constant communication with all those affected by the new technology or processes.
8) Celebrate successes at key milestones.
9) Quickly move past failures.
10) Document everything.

7.1.5.3 Post–Implementation

Once the implementation has been completed, it is important to recognize that the process does not end there. In order to ensure future support for new initiatives, there are a number steps and strategies that should be taken once a project has been implemented. The following list outlines some of the final post-implementation steps that can be taken to sufficiently wrap up the implementation.

1) Review everything.
2) Celebrate successes and review failures/issues.
3) Identify future considerations or enhancements.

4) Re-visit strategies and implementation plan and analyze effectiveness.
5) Provide a post mortem to senior management.
6) Demonstrate the working and successful solution throughout the organization.
7) Take note of those individuals who express a keen interest in what has been done.

With these checklists at the very least a good understanding of some of the more important procedures and considerations for a geomatics implementation will be covered. The authors hope that what has been provided here provides for a great starting point to build on with your own checklists as you move forward with your own geomatics implementations. We wish you every success in those endeavors and hope that our efforts to share our knowledge and experiences will benefit you and your organizations by giving you an inside look at how geomatic concepts are changing the way the pipeline industry operates.

7.2 Implementation on a Pipeline Project

Implementation on a project tends to be a lot less complex than implementing GIS throughout an entire enterprise largely due to a project's transitory nature, fewer current systems to integrate with, less data, and fewer stakeholders to satisfy. A deployment strategy would include the following items:

- Staff requirements
- Develop project Standards, Procedures and Data QC workflows;
- Create or configure the central database infrastructure;
- Check existing data based on QA/QC requirements;
- Upload or migrate existing data;
- Create metadata;
- Establish data access and procedures;
- Roll-out to project team.

7.2.1 Geomatics on an EPCM Project

Geomatics activities on an EPCM project constitute several small anticipated projects whose schedules are regularly disrupted with unexpected changes and shifting priorities (5). Rarely will you find a large pipeline EPCM project that requires only a small rigid geomatics component. Recent history has observed that as the information demands placed on the owner/operators by the regulatory process and the public at large have increased, so have the size and scope of geomatics on pipeline projects. Additionally, as geomatics has become more accepted as an efficient information asset to EPCM projects, the size of the geomatics teams on pipeline projects has grown and their scope expanded.

In the past, the intangibility of what seemed to be only short-term benefits from investments in geomatics and GIS infrastructure has delayed major GIS implementations by the owner/operators. However, as mentioned, the owner/operators have slowly accepted geomatics into their IT structures and engineering work practices. The growth of the geomatics discipline in these companies did not occur until there was an increase in computing power, the development easier to use software, a decrease in data storage costs, and general education and acceptance of geomatics.

However, with the passage of time, the long-term benefits of geomatics on the EPCM project have been realized financially and the ability to fulfill short-term exogenous demands by geomatics teams on projects has been recognized. Recognized benefits of geomatics on an EPCM project include:

- Reduced engineering cost through use of more desk top exercises and by harnessing the immense computing power of the GIS.
- A faster pace to the engineering process because of the availability of accurate, up-to-date information and the ease with which it can be communicated.
- When data is made available earlier on in the engineering process there is an increased accuracy in cost and schedule estimates.
- The fundamental ability of a GIS to communicate information over vast distances to disparate groups.
- Ability to ensure all disciplines are on the same page by maintaining a single source of truth for project data.

7.2.1.1 Structure of the EPCM Project

The organizational structure of the project team is an important component influencing the ability of the geomatics team to perform. Project structure determines the efficiency of the team as an appropriate structure reduces redundant actions, improves communication, and promotes team work. Considering the significance of the role geomatics plays on the EPCM project, geomatics has often become an independent discipline on the project in line with the Engineering, Procurement, Environment, and Construction Management teams (Figure 3).

This is equivalent to the Departmental model for large organizations in which the geomatics team is a separate department and works or coordinates with the other departments rather then having the team fall under one specific department, such as engineering and rely on the engineering leads to coordinate for the geomatics team. Experience on several major pipeline projects as well as results observed from various large organizations have demonstrated the following benefits of the Departmental model for geomatics on the EPCM project (6):

- Responds to project wide needs at a high level.
- It is able to be data focused as there is one source of control.
- Autonomy of the team allows for flexibility while also nurturing the teams own interdepartmental relationships.
- The geomatics team has a better voice in decision making and planning of data resources.
- Autonomy enables control by geomatics Subject Matter Experts (SME) over workflows and processes.

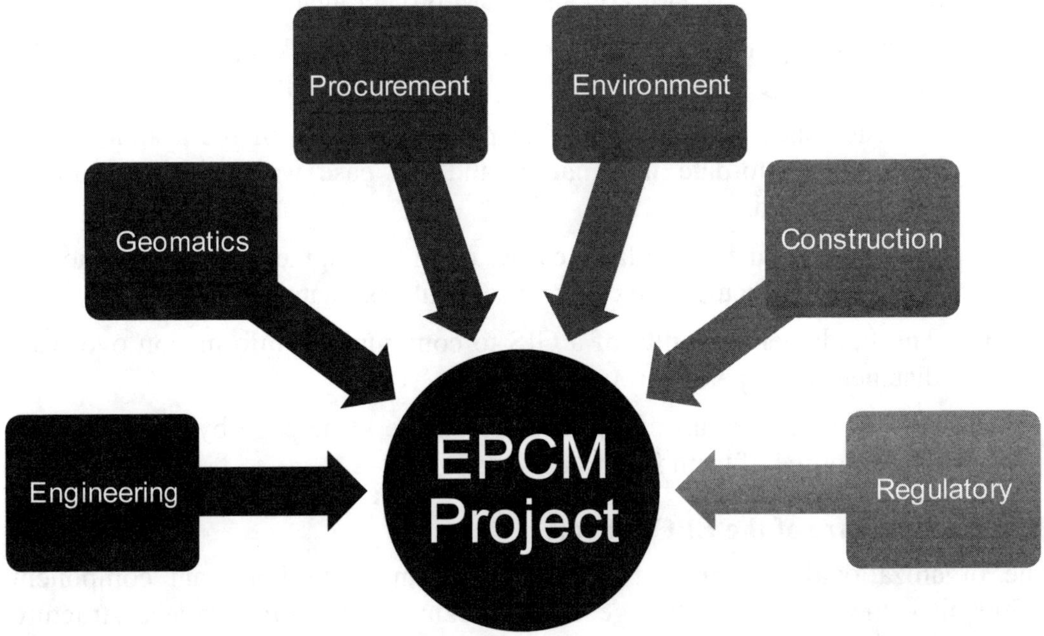

Figure 3. The Departmental Model of Organization on an EPCM project

Though the large pipeline projects could be viewed as having the Multi-Departmental model because of the presence of small GIS teams in disparate subcontracts, we do not consider these projects to be analogous to that model because the individual GIS teams are so small in size and scope they do not accurately represent that model. Further, as the lead geomatics team on a large project controls the data and the processes of dissemination, the Departmental model definition with the geomatics team as a new business unit applies.

7.2.2 Staff Requirements

There are commonly five disciplines in Geomatics that work on engineering projects:

1. Project Coordinator – Budget, schedule, task management
2. Spatial Information Systems Specialist – Database, programming
3. GIS Technician – Data management, mapping, analysis
4. Remote Sensing Technician – Aerial/satellite image processing and analysis; also includes LiDAR
5. Drafting (CAD) Technician – Drafting, automated alignment sheet touch ups

The tasks that the Geomatics team is responsible for can require contribution from each of the disciplines listed above, depending on the size of he project. As such, it is a combination of these GIS resources that sum up to the estimated effort spanning the project timeline. It is this robustness and breadth of skills that has made the application of the WorleyParsons Geomatics team to EPCM projects successful.

7.2.3 Data Management Background

Engineering data is an essential asset of pipeline owners and operators throughout the entire lifecycle of a pipeline. Hydrocarbon energy pipeline companies are increasingly generating and utilizing huge volumes of data to support their engineering and business decision-making, whether assessing potential pipeline routes, determining integrity and maintenance programs, or predicting asset lifecycles. This has given rise to the concept of *data-centric engineering*, a term which describes a modern approach to work that takes advantage of current digital technologies to collect, store, and analyze data.

Detailed records of the location of all pipeline and facility assets, as well as related land base, geographic features, environmental and socio-economic factors are all required to satisfy regulatory reporting requirements, to ensure efficient operation, and to facilitate design and construction cost savings. Ultimately however, it is not the quantity of data a company possesses that increases its competitive advantage but the quality of that data, and its accessibility to those users who transform it into valuable corporate business intelligence.

The reason we use data management practices in pipeline engineering is to ensure data integrity, security, and to manage change. Without these safeguards, projects would be open to an increasing degree of risk, directly related to the amount of data being used on a project or by an operator. As this practice becomes commonplace, competitive advantage is gained by those who effectively deal with their information assets in this way. For example, WorleyParsons has successfully employed data management on several large and small pipeline projects, which has resulted in both time and cost savings. In addition, many Owner/Operators are on the path to implementing corporate data management strategies.

7.2.4 The Central Database Concept

The notion of a centralized or Central Database (CDB) has gained acceptance as an effective way to enable engineering data management. The primary purpose of the CDB is to manage change. It does so by providing an information infrastructure for the life of the project, thus adding stability to the information management system. The structure provided by the CDB allows users of the data to conduct their activities in a stable environment, despite the constant changes being made to the data. Effectively managing data security, integrity, metadata, and change management results in a *single source of truth* that can be accessed from any part of the organization, and even remotely from field sites (Figure 4).

A rigorous CDB is composed of five main components:

1. Processes – Quality assurance, change management.
2. Standards – Data, database, survey, geodetic, metadata.
3. Infrastructure – Hardware, software, network communications, and security.
4. Data – Engineering, environmental, integrity, administrative.
5. Access – Tools (e.g. alignment sheet generator), web-based interface.

These components need to be used in combination, and are not mutually exclusive, in order to ensure effective data management. Using individual components may improve engineering projects over traditional means, but will not mitigate risks on the same order as a properly operating CDB.

7.2.4.1 Benefits of the Central Database

There are several significant benefits that can be expected as the result of implementing a CDB and spatial data management practices. First, implementing a CDB will result in the creation of a single source of data for the project. Storing all project data in a centralized location will result in increased operational efficiencies: duplication of data will be eliminated; data structures will be organized; storage space will be optimized; and, time spent searching for information will be minimized.

Second, in viewing data as a valuable asset rather than merely an expense, the proper steps shall be taken to maintain that asset, much like maintaining a physical asset. Ultimately the proper protection and maintenance of data assets will enable their full value to be realized over an extended period of time beyond the shelf life of the originally intended purpose (7).

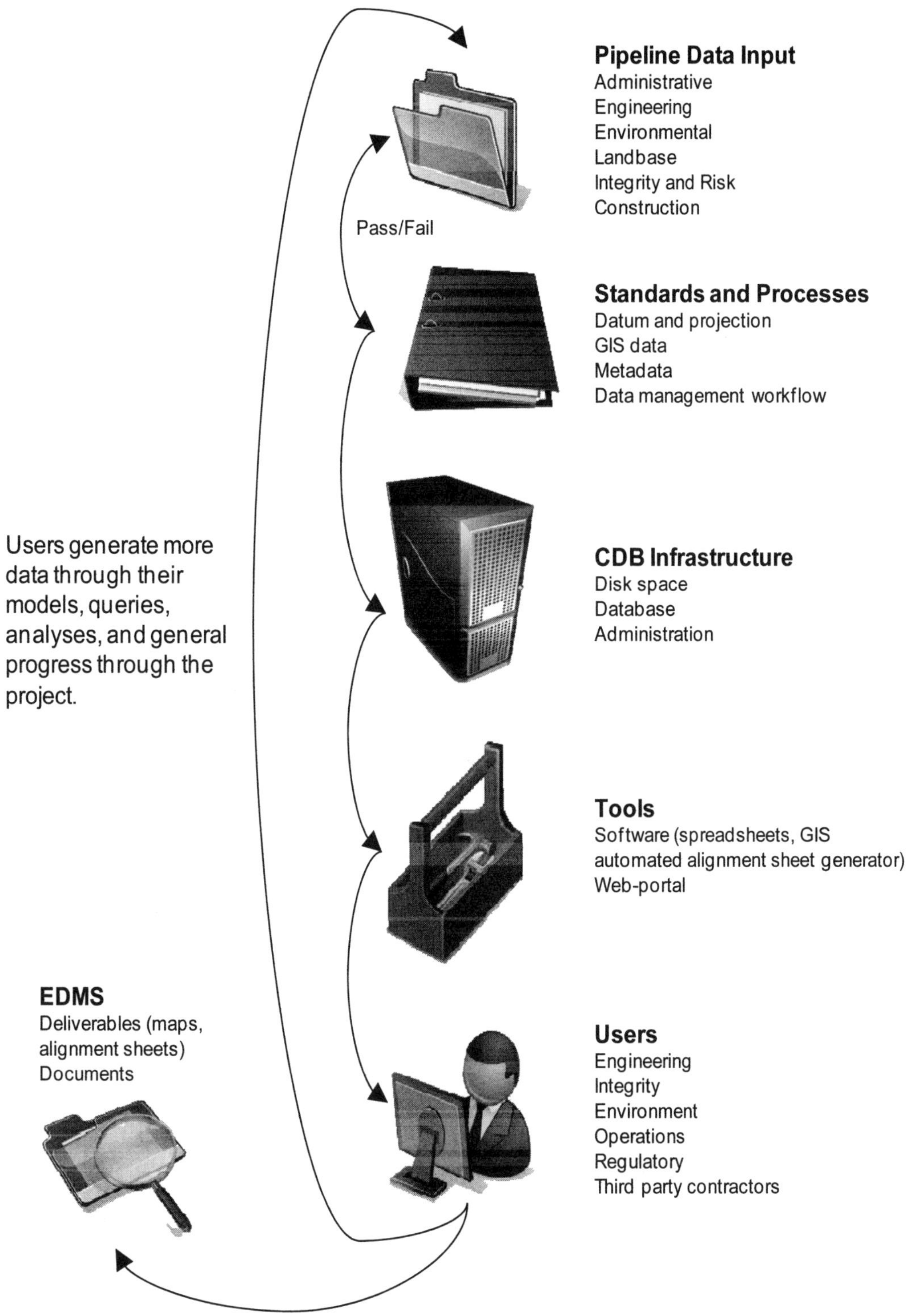

Figure 4. The major components of the Central Database Philosophy and Framework

Finally, the implementation of a CDB allows for increased control over who can edit data and release information. Maintaining this type of control through the establishment of a single source of data, user roles and permissions, and maintenance standards and processes is key to preserving the overall integrity of data (8).

7.2.4.2 GIS Support for Engineering

Geomatics supports the engineering process through detailed mapping, complex spatial and 3D analysis, and information communication. By utilizing the CDB as a primary data hub, the geomatics team is able to provide output beyond maps and alignment sheets. As demonstrated on many projects, both large (1000 + km) to relatively small (<100 km) pipeline projects, utilizing the power of a GIS with accurate, and current data yield significant information to effect the engineering process and decisions.

Many reports and services are available from skilled GIS Operators with properly managed data. The following are common services provided on larger pipeline projects:

- Routing analysis – direct support for route selection through micro-routing activities, the provision of detailed information earlier in the project then traditionally available and table top exercises;
- LiDAR or other topographical data acquisition and processing to produce surface models, profiles, or for civil works;
- Managing the collection of aerial or satellite imagery for terrain modeling, pipeline routing, mapping, and information extraction;
- Geotechnical review – providing terrain mapping and analysis (longitudinal and cross slope analysis, profiling, 3D modeling etc), Perspective analysis and inter-visibility studies;
- Environmental studies - supporting data management of field data, general mapping, field study support;
- Logistics planning – route selection and planning, mapping of routes and yards/lay down areas vehicle tracking;
- Construction planning – volumetric calculations (cut and fill), grading plans, spread mapping and calculations, resource planning and traffic planning;
- Safety – general mapping, field GPS unit support, vehicle tracking;
- Field activity support (GPS support, general mapping, survey coordination, etc); and,
- Agency Reporting – Mapping for reports and communication to government and public agencies or other stakeholders.

7.2.5 Formalize Standards Documentation

With the ubiquity of geospatial data and its growing use in pipeline projects and operations, geodetic standards are essential to maintaining reliable data. Robust standards ensure that common datum, projections, data types, and metadata establish a norm for working in a data intensive environment.

7.2.5.1 Datum Standards

Data should conform to datum standards which are widely accepted and recognized by government, industry, and the general public. Any data not belonging to one of the datum standards should undergo a transformation as specified by a given standard. Parametric data shifts should not be accepted for vertical or horizontal datum conversion. The key elements to datum standards are listed below.

In the US this could include National Geodetic Vertical Datum of 1988 (NGVD 88) and in Canada the Canadian Geodetic Vertical Datum 1928 (CGVD28). Vertical datum is always being refined, therefore, make sure one is chosen that adequately meets the needs of an organization.

The North American Datum 1983 (NAD83) shall be used as the horizontal datum. If data in the North American Datum 1927 (NAD27) is to be used, it must first undergo a datum transformation to NAD83. This transformation would be the Canadian National Transformation v2 (NTv2), or American NADCON (2.1).

7.2.5.2 Mapping Projections

The primary consideration when producing maps should be adhering to cartographic standards and guidelines. Projection parameters should suit the needs of the map and compliment the intended map use.

Universal Transverse Mercator (UTM) system is often the standard projection for data and mapping, however, the extensive geographic coverage of many pipeline systems (particularly east – west oriented) may require a custom projection such as Lambert Conformal Conic.

7.2.5.3 GIS Data Standards

The purpose of GIS data standards is to guide creating and maintaining GIS data. Effective data standards ensure data integrity and minimize confusion when new data is introduced or existing data is revised. These standards provide guidelines for data creation as well as the manner in which datasets could be validated.

The main data types encountered in any system are vector, tabular, and raster.

1. Vector Datasets

Vector datasets represent geographic objects with discrete boundaries such as roads, rivers, pipelines and land owner boundaries. Vector features are represented as points, lines, polygons or annotation. Of primary interest with vector data are:

- Format – For example, vector data shall be stored and maintained as ESRI ArcGIS feature classes or feature datasets within a geodatabase;
- Content Standards – For example, vector datasets created using geographic coordinates shall be stored in latitude and longitude and expressed in decimal degrees in proper negative values for longitude;
- Attributes – For example, attributes for shape files shall be stored in a tabular (.dbf) format, and shall adhere to attribute guidelines.

2. *Tabular Datasets*

Tabular datasets represent the descriptive data that links to geographic map feature. The file format is critical to maintain continuity among users. For example, one may only want point feature class data stored as geographic coordinates in tabular format. All other spatial data shall be stored in the appropriate vector file format.

3. *Raster Datasets*

Rasters are most commonly used for the storage of aerial or satellite imagery. However, raster datasets also represent continuous layers such as elevation, slope and aspect, vegetation, and temperature. Raster format data is typically seen as data grids or imagery, each with unique requirements

Raster image files created through satellite, aerial photography, or scanning can be stored as uncompressed TIFF or GEOTIFF files and include a TIFF world file (.tfw). However, uncompressed formats require large amounts of disk storage and occupy the greatest proportion of disk space in an enterprise. As a result, lossless compression such as ECW or MrSID helps alleviate this problem.

4. *LiDAR*

LiDAR derived data is often assessed by the accuracy required to produce topographic maps and products that meet the National Standard for Spatial Data Accuracy (NSSDA) (10).

A full-featured point cloud should be free of all data voids within the project boundary and have an average point spacing of at least 1 point per square meter. The point cloud should have a horizontal accuracy of better than 40cm RMS and a vertical accuracy of better than 15cm RMS in normal imaging conditions.

5. *Metadata*

Metadata is a cornerstone of the data management philosophy because it provides a record of change and history for any dataset compiled, stored, and maintained. Where possible, metadata should be created and maintained for all vector, tabular and raster datasets within any data structure. All metadata should adhere to the format and content standards outlined by a data manager.

Each metadata file is typically named identically to its corresponding spatial export file, and is delivered in Extensible Markup Language (.xml) format. Such a format is compatible with the output standards published by the Federal

Geographic Data Committee (FGDC; www.fgdc.gov/metadata/geospatial-metadata-standards). Standards for the creation and transfer of metadata commonly follow the FGDC's profile of ISO 19115, as described in *Content Standard for Digital Geospatial Metadata.*

7.2.5.4 Revise Standards as Needed

If any of the guiding principles outlined in each of the Standards documents is found to be insufficient or inappropriate to the project, it shall be revised, formally approved, and amended to this document. Revised Standards documentation shall be used to guide the design and configuration of the CDB, as well as all new data creation efforts.

While every effort will be made to ensure legacy files adhere to these standards, it is noted that there may be some instances where the amount of effort required to conform to these standards is prohibitive of doing so.

7.2.5.5 Summary

The adherence to established standards is critical to ensuring the accuracy, currency, and completeness of datasets in a centralized data environment and outputs generated from these datasets. Existing database standards, Geodetic standards, and GIS data standards should be reviewed to ensure they meet the needs of the project and will result in high quality datasets and project deliverable outputs.

While every effort is made to ensure new and legacy files adhere to standards, it is worth remembering that there may be some instances where the amount of effort required to create, amend, or conform to standards is extensive. Hence, any of the guiding principles outlined in standards documents should be fit for purpose and provide a framework, not a deterrent to ensure that data integrity is maintained.

7.2.6 Data Management Workflow

Many people within an organization or project will create, handle, or edit data at some time. In order to minimize data mix-ups, deletions, or unwanted changes, a data management workflow should be followed.

A central database is composed of data layers and tables held either in the *Staging* or *Production* storage areas. In the staging area, data is available for use in preliminary mapping and analysis, but is undergoing quality checks, editing or revisions. Data in the production area is considered the current version and most accurate given a specific point in time. It is production data that should be taken as the source of truth for deliverables, maps and information products to the document management system.

The main features of a standard data management workflow are (Figure 5); acquire/create datasets, staging, staging Quality Control (QC), production, production QC, rev-up data.

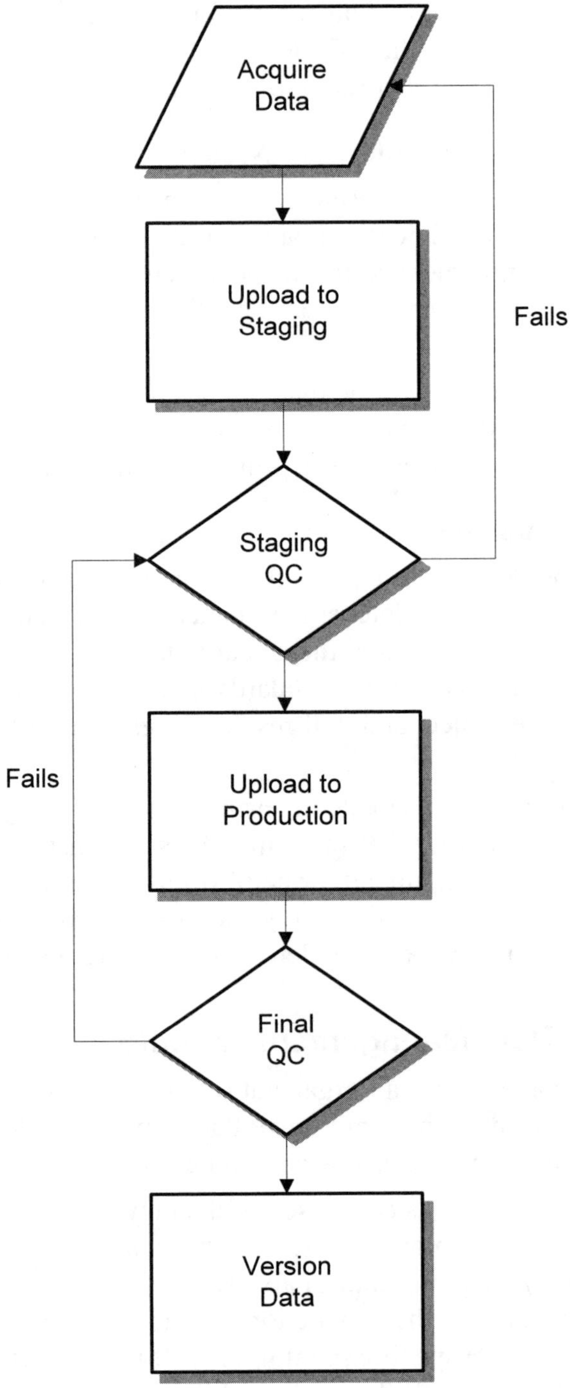

Figure 5: Data management workflow

7.2.6.1 Acquire/Create/Edit Datasets

New data is obtained either internally (through a variety of disciplines), or from an external source (third party contractor, government sources, etc.). The procedure to prepare data for uploading is the same for both internal and external data sources. Examples of datasets are discussed in the next section.

When an existing dataset needs to be modified, it typically will go through a Change Management process, which is outlined in a later section.

Once data has been acquired, created, or edited, metadata is compiled. This is an important step since metadata should be created by the data owners since they are most familiar with the data being submitted.

7.2.6.2 Staging Quality Control

When a dataset is submitted with current metadata, the Staging Administrator performs a QC check. This QC check is typically based on standards developed for the functioning GIS.

7.2.6.3 Pass or Fail

If the Staging Administrator identifies any issues with the spatial data, metadata, and/or attribute table, the dataset does not pass the initial QC and the GIS Operator, upon request, amends the dataset. Once the Staging Administrator is satisfied that the dataset is ready, the Production Administrator is notified of the datasets to be independently QC'ed and migrated to Production status.

7.2.6.4 Production Quality Control

The Staging Administrator then notifies the Production Administrator and the Production Administrator performs an independent QC of the specified datasets according to an established set of requirements. If any dataset fails the QC, comments are sent back to the Staging Administrator to be fixed. Once all the datasets pass the final quality check the datasets are migrated to the Production environment and privileges to view the datasets are granted.

7.2.6.5 Tracking

With several disparate groups using the data to produce information, centralized control is the best means to ensure all groups are serviced and that any issues amongst data provides and data users are tracked and resolved in a timely fashion. Tracking the versioning of data at the CDB level facilitates the ease and accuracy of status reporting as well as forecasting of the cross-discipline data demands and maintenance.

7.2.6.6 Versioning

The CDB allows for an ongoing management of project data through versioning and independent QA outside of the data providers environment. This ensures standards and compliance across the different groups and allows for the management of editing versioning efforts for efficiency.

Versioning results from significant changes to the data. Data is held in the Staging area of the CDB while it is being edited and maintained. Once the new version of the data is complete and has passed the QA of the CDB team it is uploaded into the Production area of the MBD and is considered a "source of truth" for the project. Versioning is reflected in standardized naming conventions.

7.2.6.7 Ownership

The purpose of the CDB is to centralize control and quality assurance, not to take ownership of the data. Individual contractors or teams, such engineering and environmental, who work with specific datasets retain the ownership of those datasets. They are also responsible for the all activities associated with maintaining those datasets (collection, value-added, quality control, versioning, analysis, metadata etc), but store their data in the CDB.

7.2.6.8 Data Information Release

During various stages of the project, agencies and stakeholders outside the project team will request access to the data used for modeling and analysis. These requests are usually to view project data or to allow the agency to conduct its own analysis without incurring the cost and time of data collection. A process to manage data information release should be established to regulate to whom, how, and when project data is discharged. The CDB can be configured to support this process through the establishment of user roles and permissions.

7.2.7 Data Composition

Centralized pipeline data sets typically fall into at least one of the categories below. Table 1 outlines some datasets encountered during pipeline design. It is important to note that one of the greatest benefits of instituting a central database early in the pipeline lifecycle is that data collected is carried forward and leveraged in future operations.

7.2.7.1 Base Mapping

Base mapping is the collection of regional maps such as government map sheets. These are used as a foundation map from which other data sources can be overlaid. Because of the ubiquitous coverage available, these maps provide information in areas where there is no data collected as of yet for a specific project. They also provide reasonably good data that can be used as a substitute when expensive data collection would not yield enough value to justify the cost. Features could include contours, hydrography, land cover, land use, roads and access, political and legal boundaries, cities, and towns.

7.2.7.2 Imagery

This includes satellite imagery and ortho-photography. Satellites have the advantage of acquiring imagery over a very large area within a single image. In addition, orbiting satellites can capture imagery in foreign countries without the need to send out a flight crew, and use parts of the spectrum not visible to the human eye to provide additional information. Although satellite data has advantages, aerial images are still relied upon to collect very high resolution images with near-survey grade accuracy.

7.2.7.3 Terrain Models and DEMs

Of critical importance to pipeline projects are models of the Earth's topographic surface, which allow for more accurate modeling of hydraulics, slopes, routing, profiling, cut and fill, line-of-sight, grade, general mobility, and others. A digital elevation model (DEM) is a raster where each pixel is encoded with an elevation value above sea level. A DEM can represent the elevation of the surface of the bare earth, or it can reveal the elevations of all features on the surface of the earth, such as vegetation, electrical lines, buildings, etc.

DEMs used in pipeline applications are typically produced through photogrammetry or light detection and ranging (LiDAR). Traditional photogrammetry uses stereo air photos to compile the elevation of ground features like break-lines and mass points. LiDAR uses a laser mounted in an aircraft to rapidly pulse a laser beam down to the earth's surface and measure the return time to determine the distance to the earth's surface. After thousands of laser pulses are processed into a raster, a high resolution and accurate representation of elevation changes along the earth's surface can clearly be seen.

7.2.7.4 Engineering Data

Engineering data consists of data relating to specific engineering details, particularly any event which is near, on, or crosses a pipeline. Common types of engineering data can include items such as the centerline, pipe material, coatings, welds, crossings, profile, in-line inspection data, digs and repairs, cathodic protection, assemblies, right-of-way, and workspaces.

7.2.7.5 Environmental

Data in this category is related to environmental factors which are not engineering specific, and includes data regarding the use of environmental resources. Common types of environmental data include aquatics, fish and fish habitat, water quality, groundwater, vegetation, landforms, permafrost, air quality, wildlife, historic resources, land and resource use.

7.2.7.6 Administrative

This category contains data associated with government or survey activities such as municipal and provincial boundaries, private lands definition, registered plans, land owners, mineral rights etc.

7.2.7.7 Questions to Ask

When acquiring GIS data it is important to consider several issues.

- If the data desired is not readily available, how long will it take to produce?
- What is the quality of the data in terms of the positional accuracy (location), and attribute accuracy (ex if a polygon shows an area having a population density of X, how close is the actual population density to that value)?
- Are data sets being combined from different sources (ex two adjoining areas surveyed for soil properties by two different companies), how will they align/match up to one another?
- How current is the data, is the data too old to be useful?
- What is the financial cost of obtaining the data?
- Will the data be delivered with a file size and in a file format that is usable and convenient?
- Not all GIS data can be freely shared or copied within an organization, or between companies. What are the licensing/legal agreements associated with the data?

TABLE 1: TYPICAL DATASETS FOR PIPELINE DESIGN

Data Descriptor	Data Example
Pipeline Facilities	Pipe information, external coating, internal coating, valve information, casing, launcher receiver information, pipe bend, flange, fabrication,
Geographic Features	Roads, access, foreign line crossing, railroad, land use, right-of-way, workspaces, hydrography, land owner, cadastral
Location	GPS point, monument, profile, centerline geometry
Raster/Imagery	Scanned mapsheets, aerial/satellite imagery, LiDAR, cross/longitudinal slope grids

Environment	Aquatics, fish habitat, water quality, groundwater, vegetation, landforms, geotechnical, air quality, wildlife, historic resources, land and resource use
Stationed Centerline	Line, route

7.2.8 Data Quality

An essential part of any GIS functioning as expected is quality data. With the amount of digital data available today, combined with new and traditional ways of developing digital data, it is critical to have a common means to identify and describe quality data. To assist the identification of quality data a set of GIS data standards should be prepared as well as a Quality Control/Assurance procedure that will ensure that only high-level quality data is accepted and introduced into the GIS database.

When defining GIS data quality the following prime components of the GIS data should be reviewed and accuracy standards applied to them:

- Spatial Placement;
- Connectivity;
- Database Design Conformance;
- Database Attributes;
- Age of Data;
- Data Completeness.

7.2.8.1 Data Quality Validation Guidelines

- When the spatial accuracy and connectivity of the data is validated the following should be considered:

  - Use of correct spatial reference system;

  - Data should be free of topological errors;

  - Data format used should be consistent within the database.

- Geometric networks should be used to validate the GIS data connectivity, such as: object, database and device connectivity, assure the GIS database accurately models the real life network system. For database design conformance and database attributes validation the following should be considered:

  - Using a suitable Data model standard will provide easy assessment of existing data in the database and the ability to apply standard business rules for a multiple datasets.

- The age of data has a direct correlation to the source used to provide the data. In order to assure that the GIS system users make the appropriate decisions every effort must be made to use only the most recent data and properly document the time data was collected.

- For the completeness of the GIS data it is critical that the database files capture all objects of the automated pipeline system, such as number of compressor stations, meter stations, valves, pig traps. Also, generalization, omission, selection criteria and other rules used to define datasets must be properly explained and a clear definition of all possible error specifications has to be provided.
- Creation of proper metadata and establishment of metadata standards or adoption of existing metadata standard is another important aspect of creating and maintaining quality GIS data.

7.2.8.2 Benefits of using quality GIS data for pipeline applications:

- Quality integrated spatial data from GIS systems and real-time data from SCADA can be used as a decision making tool by relating the pipeline data collected at a specific location (meter station, pump station) to changes in the data over time.
- Assures efficient emergency response planning, by providing the added functionality of a GIS such as use of spatial queries to assess and evaluate existing infrastructure, demographic data and available access routes relevant to specific pipeline infrastructure.
- Accurate 2-D and 3-D graphic visualization and modeling of pipeline systems and operating devices.
- Provides an efficient and cost-effective creation of digital data ready for integration with SCADA of detailed as-built plans for meter station, pump station, pipelines.
- Effective pipeline integrity work documentation and maintenance planning (section 9.3.2 Integrity Tools).

7.2.9 Change Management

The main reason for spatial data management is essentially to manage change. This occurs by providing an information infrastructure for the life cycle of the pipeline and adds stability to an information management system. A certainty in this digital world is that the amount of data required and generated is immense and continues to grow.

Change Management means accountability. For an organization, change management means defining and implementing procedures and processes designed to deal with changes to any item that should fall under some form of version control.

Change management provides a paper trail giving an accurate history of any changes applied, the reason for the change, and of course pre-approval and post approval signatures. Change Management must be a structured process resulting in the validation of the proposed changes as well as an accurate status reporting tool for all affected parties.

7.2.9.1 Controlled Change Management

The role of Change Management is to add organization to the potential for disorganization, left long enough uncontrolled change can become a logistical nightmare. It plays an important role in any organization since the task of managing change is not an easy one. Successful change management should involve selling the process internally and creating an efficient infrastructure for change management.

Managing change requires a broad skill set, political, analytical and people skills. These skills are important during and after the initial implementation. The implementation of Change Management process demands a whole new way of working in an organization, and also involves looking differently at other processes that may be affected.

A Change Management system is generally tailored to suite the needs of a Project or Company, but should also be adaptable to any project within the organization.

It is important to consider the two user groups when dealing with controlled change, the *requestors* and the *implementation team*. Left uncontrolled the potential for multiple changes by multiple requestors can often lead to confusion on the implementation side and issues with the integrity of the changes. Through control and approvals only the requested operation should be implemented with little room for after the fact amendments.

7.2.9.2 Proactive not Reactive Change Control

Identifying the need for a Change Management system should be a priority in the initial stages of a project, playing catch up at a later stage can introduce integrity and revision issues.

Implementing controlled change in the initial stages of a project will help reduce the resistance to change as most people have negative attitudes and perceptions towards change. If the process is implemented as part of the ramp up it will almost always be seen as part of the project procedure.

Guidelines to Implementing and sustaining a Change Control process
The potential for resistance must be taken into consideration, and why that resistance may occur. The potential for resistance means implementation of a Change Management process will involve political, analytical and people skills.

- Know your data structure and understand what data must be controlled.
- When possible assign ownership.

- Implement the process in steps, so as to minimize the overall impact on current structures and processes by, identifying the Change requirements through data collection and analysis.
- Minimize the administrative traffic jam by making the process user friendly

Spatial data management provides organization and structure to the complex web of information demands for a project or organization. The structure it provides allows data users to conduct their activities in a stable environment despite the constant changes being made to the data. With a dependable information system backbone, data use and sharing become valuable additions to workflow.

7.2.10 The Database and Data Models

Given the vast amounts of data collected and stored for pipelines, such project data needs to be properly managed so as to ensure that it can be easily accessed, retrieved, and utilized within the GIS framework. This is most effectively achieved with an expert Spatial Data Management System (Figure 6), providing a database-centric configuration complimented with an industry-standard pipeline data model. Leveraging the database and data model is typically a customized front-end application suite, which is responsible for data management, analyses, and reporting activities, all of which help compose the larger GIS context.

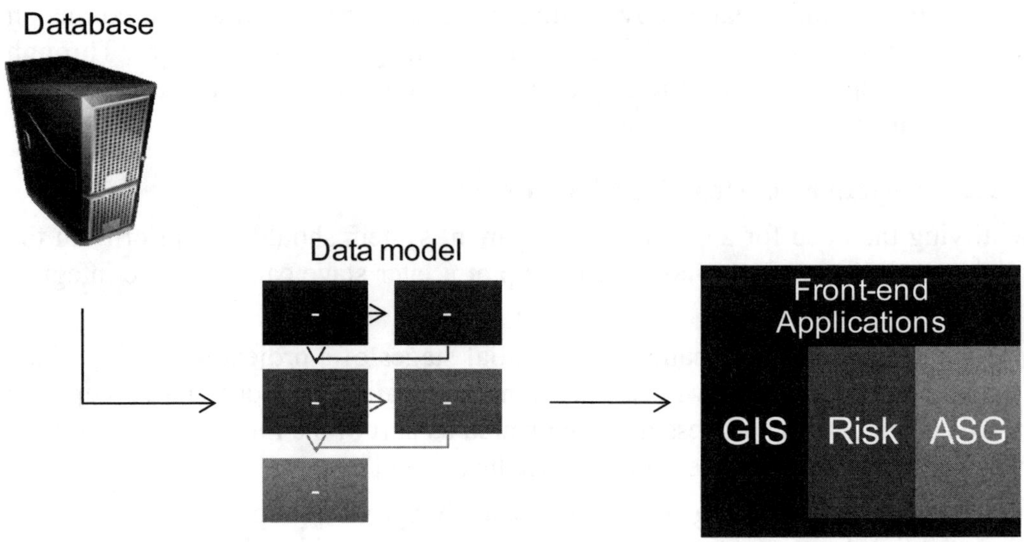

Figure 6: Spatial Data Management System

7.2.10.1 Database

The database is a major component of a Spatial Data Management System as it serves as a warehouse for project data. This means that the database can store both spatial and non-spatial data. Spatial data includes base maps, satellite imagery, aerial photographs, digital elevation models, pipeline centerline, cadastral layers, and any other data associated with geographic locations. Non-spatial data includes engineering reports and drawings, administrative documents, pipeline assets and operations information, stakeholder information, and other business information pertaining to the project. Since project data is being utilized by many disciplines within an organization, the problem of having data 'accidentally' modified from the original or removed from its intended storage location is inevitable. As a result, the integrity of the data is lost and time will be spent to investigate or rectify the problem unless a proper database is employed.

Since pipeline data continues to accumulate throughout the life of the pipeline, the underlying database of a Spatial Data Management System for any pipeline project should always be built with sufficient capability to house current project data inventory and support future project growth.

7.2.10.2 Data Models

Over the last decade, pipeline operators have formed associations to develop industry-standard data models for managing gas and liquid pipeline assets, as well as operations information. These pipeline data models have served to enhance pipeline companies' business operations in terms of cost, effectiveness, efficiency, and reliability. Benefits of these pipeline data models include:

- Elimination of the need to develop a data model from scratch;
- Reduction in integration and implementation time for new software applications;
- Adoption of proper data management techniques resulting in improvement of effective business processes; and
- Customization of existing data model to meet specific project needs.

There are several pipeline data models available in the pipeline industry; however, ISAT, PODS (Pipeline Open Data Standard), and APDM (ArcGIS Pipeline Data Model) are the most widely used.

1. ISAT Data Model

The ISAT data model was created as an open, publicly-available, standard data model for the pipeline industry. The model provides support for all the basic facilities routinely managed by a pipeline company along with operational data required by federal regulatory agencies. ISAT is database independent and supports the following activities: facility data maintenance, as-builting, alignment sheet generation, risk assessment, field data collection, pipe inspection, integrity management, web-based data, reporting and analysis; and, analyst tools including class location analysis, MAOP calculations and marketing. Many organizations have implemented a modified ISAT model to meet their unique business needs.

2. PODS

PODS (Pipeline Open Database Standard) is developed and managed by the PODS Association, which is a non-profit pipeline trade association. As a popular database model in pipeline data management, it is a completely open data standard. PODS provides a general model by which we can create a standard pipeline GIS database, and is easy to expand by adding different aspects as required.

The objectives of PODS include:

- A standard data model to minimize data design activities;
- Reduce database development risk;
- Easy to expand and maintain;
- Provide an integration environment with multi GIS software platforms;
- Exchange and share a standard database with partners;
- Formalize and optimize data relationships;
- Developed in RDBMS such as Oracle, Microsoft SQL Server, and MS Access.

PODS is widely used in the oil and gas pipeline industry, and many companies currently use this standard data model to manage their data. Furthermore, many commercial pipeline applications (such as automated alignment sheet generators) are using PODS as the backend data model for their operation.

3. Customized APDM

The ArcGIS Pipeline Data Model (APDM) was initially derived from both the ISAT and PODS data models and is expressly designed for implementation as an ESRI geodatabase for use with ESRI's ArcGIS and ArcSDE products. The APDM was designed as a template that could be customized by operators to account for their unique processes and assets, with the primary objective of incorporating tools for linear referencing.

7.2.10.3 Front-end Applications

The database and data models form the foundations for the Spatial Data Management System. However, data cannot be adequately leveraged without any form of interpretations. End-users are typically not interested in how the data is stored, but rather much more interested in what the data represents. For instance, a DEM stored in the database has little meaning if the user has no way of visualizing it to understand the terrain representation. Once the user has an understanding of the terrain, pipeline micro-routing or hydraulic studies can be electronically performed with the availability of the appropriate customized GIS tools and other project data residing in the database. In order to assess the integrity of the pipelines to detect potential issues and plan for remediation, inline inspection data needs to be analyzed and represented in the forms of charts, reports, and geographical visualizations. It is the representation of the stored data that allows end-users to make informative decisions to better serve their business operations in terms of cost, effectiveness, efficiency, and reliability.

Since many pipeline companies have increasingly adopted the implementation of industry-standard pipeline data models, pipeline engineering software vendors have begun to use these as their underlying data models for application development. As a result, integration to a variety of readily available software is seamless when the foundations of the Spatial Data Management System are in place.

An expert Spatial Data Management System is one that leverages the technologies and data models to make available front-end applications for data management, analysis, and visualization activities.

7.2.11 GIS Infrastructure

A high-quality Spatial Data Management System requires a robust infrastructure to ensure data integrity and reliability in its use. A common configuration is shown in Figure 7. The main components of this arrangement are; database server, GIS server, web server, and workstations.

7.2.11.1 Database Server

The database server of a Spatial Data Management System has similar behaviors as any client-server architecture. This means that the database server stores data and processes, and responds to data requests sent by any client machine running front-end applications.

Since the database server is a central system, it has the ability to enforce strict security and control access by allowing only authorized personnel to update the data. As a result, data integrity is always maintained. In addition, the client-server architecture is versatile such that changes or upgrades performed in one environment do not disrupt the services of any other. This is a great advantage, as computer system down time should be an avoidable expense for all business operations.

7.2.11.2 GIS Server

The GIS server is used to deliver GIS capabilities to client machines and it behaves in a similar manner as any client-server architecture. This means that GIS requests are processed on the server side and results are delivered to the client machines in a very efficient and timely manner. This is an advantage, as the client machines are not bogged down with processing heavy-load GIS tasks. The GIS server can also deliver maps and GIS functionalities to the Internet.

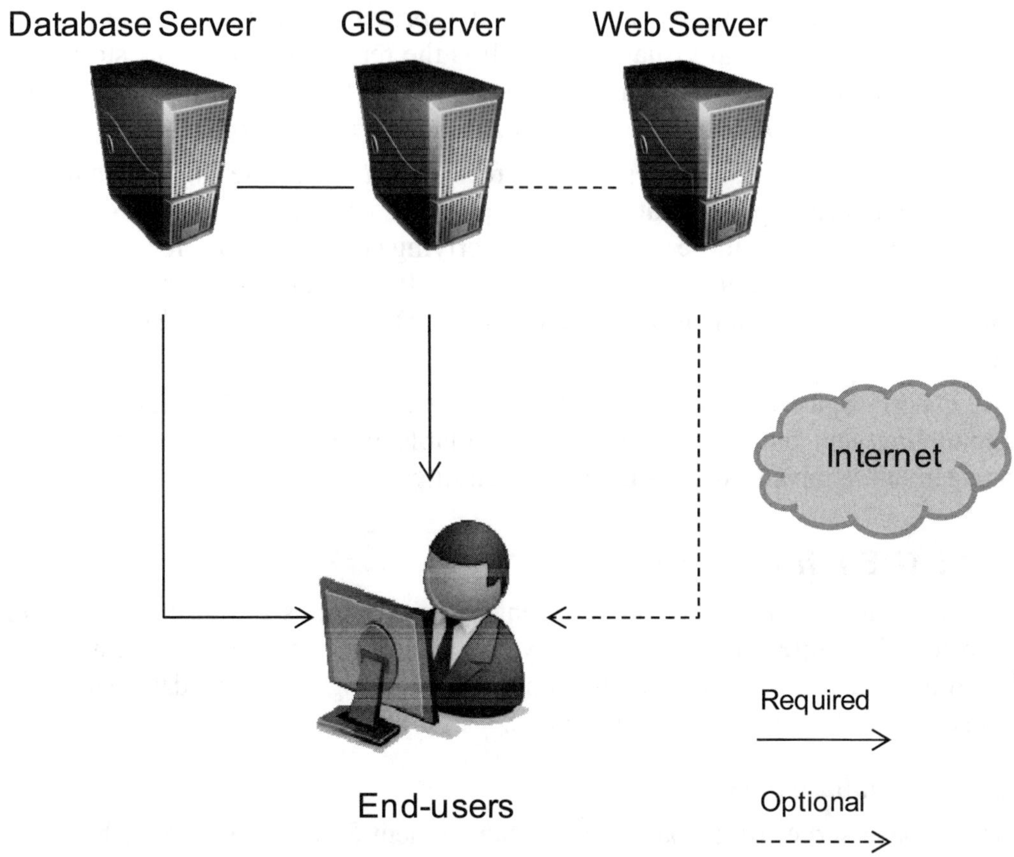

Figure 7: GIS Infrastructure

7.2.11.3 Web Server

With the convenience of the World Wide Web, we are able to access almost any information in any format at the click of a mouse button from anywhere in the world. In the context of a Spatial Data Management System, the web server is responsible for rendering front-end web-based applications to web browsers on client machines.

In general, web applications are in the forms of HTML or ASP pages containing scripts, macros, and multimedia files. Web servers then serve these pages using HTTP (Hypertext Transfer Protocol), a protocol designed to send files to web browsers.

7.2.11.4 Development Workstation

A development workstation is a client machine for software developers to develop and test the Spatial Data Management System. This is where various software, development platforms, and software developer kits are installed to allow software developers to migrate data into the database and create specific programs or applications.

7.2.11.5 End-user Workstation

This is where front-end applications are installed and available for end users. If third-party tools are used in creating the front-end applications, then this client environment must have the necessary runtime objects or desktop applications of those tools to be able to access the desired functionalities. If the front-end applications are web-based, then the client machine must have a web browser, internet connection, and be configured to accept scripts, applets or other specific criteria that the web applications require.

7.2.12 Prepare Infrastructure

The purpose of this procedure is to describe the work required to transition from an existing environment to a structured central database environment that supports spatial data management practices. Implementation requirements are presented below in the chronological sequence that they should be performed.

7.2.12.1 Assess Existing Disk Storage

The first task is to locate and inventory all project data. It should be noted that multiple data sources and locations often exist. Once all project data has been located and inventoried, disk space allocation should be assessed. Taking into consideration possible areas of duplicate data, as well as storage requirements for future data acquisitions (such as new imagery), an estimate should be made of total disk space required.

7.2.12.2 Assess Existing Software

A review of the current software used should be conducted, including an assessment of RDBMS, GIS, and electronic document management applications and tools. It is possible that existing software applications will meet project needs. However, in the event that a potential deficiency is identified, the necessary steps should be undertaken to procure additional software to satisfy the gap.

7.2.12.3 Migrate Data

1. Identify Data for Migration

An inventory of all project related data and information should be compiled. All duplicate datasets should be eliminated, or versioned as appropriate. Spatial and tabular datasets should be separated from project deliverable outputs (i.e. documents, drawings, maps, and spreadsheets). Spatial and tabular data that represents the single source of truth should be prepared for migration to the CDB, while project deliverable outputs are prepared for storage in the Electronic Document Management System (EDMS).

2. *Classify Data into Categories*

Each spatial and tabular dataset should be organized into one of the six categories, noting that Imagery and Terrain/DEM data may be stored separately from the vector data, outside of the geodatabase:

- Base Mapping;
- Imagery;
- Terrain/DEM;
- Engineering;
- Environmental; and,
- Administrative.

There may be a small number of exceptions where a given dataset cannot logically be assigned to one of the above categories. In such cases, the dataset may be assigned to a seventh feature dataset category to be called *Miscellaneous* or *General*.

3. *Perform QA/QC*

All project data must adhere to the standards set forth in the appendices of this document and will undergo Quality Assurance and Quality Control (QA/QC) checks to ensure the integrity of the data before it is migrated to the CDB.

All databases being included in the CDB must abide by the format guidelines within this document and shall be named in accordance to the naming conventions.

Spatial data will be tested for positional accuracy and must adhere to the Geodetic Standards. All GIS data created as part of the project must adhere to data creation standards.

4. *Create Metadata*

Metadata will be generated for all new datasets created for the project. Metadata will adhere to the project standards. Metadata are often documented in .xml file format and named identically to the individual file it represents.

While every effort will be made to ensure metadata is generated for existing datasets, it is noted that there may be some instances where the amount of effort required to generate complete metadata is prohibitive of doing so. In such instances, an attempt will be made to document at least basic information such as: source, date of creation, date of last update, etc.

5. Migrate Data to CDB Production Area

Once existing spatial and tabular datasets have been identified as the single source of truth (not a duplicate dataset), have undergone QA/QC checks, and have had the appropriate metadata generated, the dataset can be migrated to the appropriate storage location within the CDB. Datasets will be stored as feature data classes within one of the established feature dataset categories.

It is critical that file names adhere to the file naming conventions outlined in project standards.

6. Migrate Project Deliverables to EDMS

All other project deliverable outputs such as alignment sheets, maps, drawings, and reports should be migrated and stored in the EDMS.

7. Testing

Upon the completion of migrating data to the CDB and to the EDMS, testing should be undertaken by the project team to ensure access to project data and information is available to users as intended.

7.2.13 Access to the Central Database

We often feel good about having secure data with spatial integrity but really need to get access to it in order to really leverage the GIS. Secure and authorized access to data stored in the CDB can be provided through the following methods:

7.2.13.1 GIS Desktop Suite:

This is software such as ESRI and other GIS packages that perform data management, mapping, and analysis functions. This software is typically used by GIS technicians to support engineers and other team members.

7.2.13.2 Custom Applications:

Based on project needs, an IS team may develop custom applications that can access, and in some cases update, project data. Such custom applications may include a Field Survey Application, a Crossings Database and/or a Compliance Management System.

7.2.13.3 Web Portal:

The inherent nature of the Internet is to allow the sharing of data. Since a Web GIS can be made available through the Internet, it provides the facility for the sharing and publishing of data. This proves extremely beneficial for the collaboration of groups not physically working together and communicating with people in the field. Since the architecture of a Web GIS follows a heavily centralized model, it offers many advantages:

- Rapid System Deployment. By using a Web GIS model, a system can be modified in one place (i.e. the server), and subsequently, all users will always be accessing the latest version of a system, without having to perform software upgrades/installations.
- Unrestricted Deliverability. Traditional computer systems are often difficult to make accessible to users outside of a particular area, whether it be a geographical area (i.e. must be in the same building, same city) or computer network.
- Support for Different Clients. Web GIS requires that the end user have only a basic web browser to use it. This makes it possible for users of different operating systems and hardware platforms to easily access a Web GIS.
- Data Sharing. The inherent nature of the Internet is to allow the sharing of data. Since a Web GIS can be made available to the Internet, it provides the facility for the sharing and publishing of data.
- Systems Integration. A Web GIS is in itself part of a Service Oriented Architecture (SOA). That is, it provides a service, being GIS functionality, to a variety of clients, including other systems. Given this, it lends itself to being easily integrated with other server-based services, for example, a SCADA system could provide real-time data that could integrated with the Web GIS and displayed on a map.

7.2.14 Implementation Roll-Out

7.2.14.1 Communicate New Architecture to Project Team

The project team should, of course, be informed of the development of the CDB structure and new data management practices from an early stage, and should be involved in the implementation process as appropriate. At the time when each Standards document is finalized, it should be provided to all team members who are involved in data creation and maintenance activities. This will likely be prior to the 'go live' date for the CDB, to ensure that all new data being created adheres to the established standards.

Once all project data has been migrated to the CDB and the EDMS, a demonstration should be provided to all appropriate team members. The demonstration should provide a clear overview of the CDB structure and file locations, in addition to discussing at a very high level the standards that have been implemented. Team members should be provided with the contact information for the person responsible for the management of the CDB.

7.2.14.2 Maintain Standards Documentation

Finally, it is important to maintain the Standards documentation for the project as a living document. A process should be in place to ensure that any changes to Database, Geodetic, or GIS Data standards are captured and revised in the appropriate documentation. This process should also include a method for informing the appropriate team members of the revisions.

7.3 Some Common Implementation Issues

As with any software or hardware implementation, there are fears of endless spending with little progress and other pitfalls. We also have been strong advocates throughout this book of using spatial data management systems, therefore, below are some of the most common issues encountered in geomatic implementations, they are presented as an issue with corresponding approaches.

Issue #1: There is limited support or interest in your data management service or geomatic tools from users or management.

Approach #1: Improve communication and participation in project planning. Be inclusive, do not forget to include non-spatial components like CAD and EDMS in your discussion. Do not operate out of schedule-based fear, being inclusive does not take long.

Approach #2: Do not lose sight of your customer focus. Your customer can be your client and an also be users within your own project team. Remember, spatial data management systems are for the users, to make their jobs easier with less risk and greater productivity. These systems are not just a data repository. Make sure your business requirements have really come directly from the customer and users.

Approach #3: Make sure your implementation and production teams are strong and responsive. This is related to Approach #1, where effective leadership and strong foresight will provide a team with a secure sense of purpose and participation by all invested parties.

Issue #2: Users are not receiving adequate support to their queries and needs.

Approach #1: Users should get involved with the geomatics team. Do not let biases and technology fears impede, what can be, a very successful addition to pipeline engineering. Make your needs known and collaborate, even if it takes a little longer. Like any specialty team, a geomatics group can get isolated and lose customer focus. This tends to take the form of data management tunnel vision; where the geomatics group excels at managing spatial data at the expense of providing efficient access to all the data being managed.

Issue #3: An overwhelming feeling that a project or organization is too far along to implement GIS.

Approach #1: Do the implementation. These feelings that question the feasibility of an implementation are based on fear, not the objective evidence that shows GIS improves pipeline projects (ref). We don't question whether spreadsheets make our lives easier, GIS is no different. The only consideration is how large to implement; Approach #2 will deal with this.

Approach #2: Make sure your implementation and production teams are strong and responsive. They will know how involved an implementation needs to be and be able to do it quickly while still being inclusive in the process.

Issue #4: The implementation is not progressing on budget or on schedule.

Approach #1: Contrary to what is expected, this will happen because people are in a hurry. The geomatics project manager needs to be deliberate and methodical in their understanding of the client's and users needs. In addition, all invested parties need to collaborate to create a success. A slow implementation is not just the implementer's immediate problem. If the users and client do not get involved, it is eventually their problem.

Approach #2: Similar to #1, over-budget comes from groups not working together. Stronger collaboration will involve greater communication and build understanding and trust.

Issue #5: An implementation is near or at completion and users still don't understand what to do.

Approach #1: Communication and user-training is very important. Geomatics professionals are proud of their work, it should be shared, enjoyed, and used to its fullest potential.

Approach #2: Make sure standards are shared and workflows are known. These are shared expectations, so they should be spread widely. This is also particularly important when the system is in production since team members and vendors must interface in a common language, that is the language of the standards.

7.4 What's Ahead?

This chapter has some detail and descriptions that are quite involved and can be overwhelming. Of course, for experienced geomatics professionals, an implementation is challenging, but definitely not a fearsome undertaking. Perhaps some of the more notable advances we expect to see in the area of GIS implementations would be:

- Fast, simple implementations that would leverage pre-built standards, processes, and workflows, and existing data management environments.
- Web portals with fast refresh for imagery or other data types so users anywhere are able to query project data without the need for specialized software.
- Image compression and storage technologies that maintain image quality while delivering fast uploads and display as well as minimizing hard drive space requirements.
- More open peer-to-peer style data management systems with automated

QA/QC so data can be accessed from numerous sites, but still maintained with the rigor of local data management.

References

(1) Delong, Thomas J. & Vijayaraghavan, Vineeta, "Lets Hear it for B Players", Harvard Business Review, June 2003

(2) Pearlson, Keri E., Saunders, Carol S., "Managing & Using Information Systems, A Strategic Approach", P. 307, John Wiley & Sons Inc., Hoboken, NJ, 2006

(3) Hartman, Francis T., "Don't Park Your Brain Outside", P. 25, Project Management Institute, Inc., 2000

(4) Tomlinson, Roger, "Thinking about GIS Geographic Information System Planning For Managers", P. 31, ESRI Press, Redlands, California, 2003

(5) Hlady, J., Buchanan, K., "Success and Failure in Pipeline GIS – Lessons Learned", Proceedings of International Pipeline Conference. Calgary, Canada. IPC2006-10500, 2006

(6) Lemire, Pierre, 2003. "GIS Organizational Structures," http://www.urisabc.org/assets/events/2003/MovingForward/presentations/Speakers/1LeMire/pres1_LeMire.pdf

(7) Adam, S. and Hlady, J., "Data is an Asset that Should Be Managed." Proceedings of International Pipeline Conference. Calgary, Canada. IPC2006-10500, 2006

(8) Adam, S., "A Central Database Improves Pipeline Projects." Proceedings of International Pipeline Conference. Calgary, Canada. IPC2006-10373, 2006

(9) Refer to http://www.fgdc.gov/standards/projects/FGDC-standards-projects/accuracy/part3/chapter3

8 Industry and Technology Trends

As with any discipline, both pipelining and geomatics are constantly evolving. Changes are motivated by regulatory and technological advances as well as public opinion and perception. These changes, combined with the natural evolution of geomatics and the business of pipelines, result in the emergence of new practices and processes. Through this emergence, innovations are made upturning previous processes and practices and materializing new paradigms. This final chapter provides a discussion on current trends that are gaining momentum and offers some insight into how these trends are having an impact within the pipeline business.

8.1 Worldwide Infrastructure Development

As global population grows, so too does the need for infrastructure to support us. Evidence of this can be found all around us, from aging roads and bridges to outdated schools and hospitals. The story is much the same no matter where you go on the planet. Even if the population growth were to stagnate, the need for more infrastructure on a global scale would still be evident. Essentially, in both developed and developing nations an increasing gap is emerging between the need for infrastructure and the rate at which new infrastructure is being built (1).

Governments have historically taken on the responsibility of providing the infrastructure required by the societies that they govern. The ever increasing infrastructure gap that the world is facing has now become large enough that most governments do not have the means to effectively mitigate the problem. Increasingly, public-private partnerships (PPPs) are being recognized by governments as an effective strategy for closing the infrastructure gap. The main benefit of PPPs is that through the involvement of the private sector the lifecycle for building new infrastructure including financing, design, construction and operations becomes much shorter. The implication of this is that in the coming years more infrastructure will be built at a faster pace than has occurred in the past. Involving private business in these projects will, over a much shorter time frame than could be accomplished by government alone, serve to reduce the infrastructure gap and in doing so increase the quality and standard of living around the globe.

How does this relate to pipelines and Geomatics? There is a very real and direct connection. The energy required to both build and support new infrastructure will, in many cases, be delivered via pipelines. Furthermore, with new infrastructure there will also be an ongoing need for utilities to support that infrastructure and the societies that utilize it. Thus pipelines will play a major role in both the construction of new infrastructure and also in supporting it as a going concern after construction. The use of Geomatics within the pipeline life cycle will serve to make the role of pipelines within infrastructure development a more efficient one.

Moreover, pipelines are infrastructure items themselves. In the developed nations of the world, pipelines are mainly built, owned and operated by public corporations and private companies, as opposed to government. In many countries however, pipelines are infrastructure items that fall under the responsibility of government, especially in nations that have National Oil Companies (NOCs). Under this scenario, PPPs are likely candidates as an effective means to develop much needed new pipelines. Under a PPP pipeline infrastructure development scenario, the use of Geomatics within the pipeline lifecycle will be driven by two main factors; the efficiencies brought forth by the private sector participants and the need for tools that both groups can leverage to support cooperation and effective communication with one another.

The following sections discuss emerging geomatic technologies and concepts and their effect on the pipeline business. Many of these have the potential to substantially change the way the geomatics discipline interacts with other disciplines. The pipeline discipline will feel the effects of these new technologies and processes and hopefully be able to experience positive change and new paradigms itself as a result.

8.2 Virtual Earth

The virtual Earth is one of the most exciting geomatic platforms in recent times, not just to pipelining, but also to the general public. Essentially, the virtual Earth is the new generation of world atlas where a user can sit at a computer and travel the world just by moving a mouse. With current road networks, satellite images, and business and street addresses, it has truly made linking the *what* and the *where* easier than ever before. Internet applications like Google Earth, Microsoft's Virtual Earth, and even earlier web applications like MapQuest have brought the corners of the world closer to us with realistic detail rarely experienced before in the past.

A digital virtual Earth consists of particular geo-spatial layers of interest that are cached and then served over the internet. A user can locate places and objects of interest and do simple queries to get address information and basic attributes. In addition, users can upload data to be overlaid onto their virtual map and, if desired, viewed by others across the internet. There are also stand-alone applications that do similar things but require the user to have and maintain their own data linkages.

8.2.1 How It Is Used

Pipeline professionals have taken to this technology quickly since it provides proximity information to anyone with a computer and internet connection. Remember, that since a pipeline is a geo-spatial phenomena, we are constantly looking at *what* and *where* questions. What used to require buying a paper map and drawing on it or possibly even taking a field trip is now done at the desktop with a high resolution satellite image and other layers like roads and contours. A typical process would be nearly identical to how it is done with pens and a map, only digitally:

- A pipeline engineer has a start point A and end point B
- She defines places along the route the pipeline needs to go, like facilities, and areas to avoid
- A first revision route can be digitized on the screen, avoiding features, crossings, and landscapes visible in the data being viewed
- This route can be saved (like a .kml file) and shared

Tight data management controls are required so that pipeline route revisions are not confused. One thing the digital world has done in making things so easy to do (like create dozens of route within hours) is created volumes of data that need to be managed. Please take care to do so to avoid any confusion later in the design process.

8.2.2 Benefits

As with most innovations, some benefits appear quickly and are obvious. Others may be subtle but can still be significant.

- The one theme that appears through many of these applications is the ability to do work at the desktop. Desktop studies do not to replace field work but minimize the need for fieldwork from a cost, safety, accessibility, environmental disturbance, and other perspectives.
- Using a digital Earth is very effective during conceptual engineering or other phases where budgets for data collection and field work are restricted
- An engineer is able to complete numerous routing scenarios quickly.
- Information sharing and collaboration. More of this is discussed in the next section but data layers, photos, and other information which are location dependent can be shared across the internet using sites like Google Earth
- Learning and using the applications (web-based or stand-alone) is very simple

8.2.3 Considerations

With virtual Earth systems, there are three main considerations when planning to use them and understanding how they affect your project:

- Positional accuracy – Since there is not a lot of time spent correcting data on these sites, the geo-spatial accuracy must be given a wide margin of error. This is difficult to accurately quantify but errors of about 20m in low relief areas are common and can be greater than 100m in areas of high relief. The point here is not to discourage the user, but to inform the user that routing a pipeline through a high density area or sensitive area will require more accurate data in the future as engineering progresses.

- Data availability – Typically, we prefer or are required to do desktop studies because areas of interest are inaccessible or remote. However, these areas are sometimes the same ones with the least data coverage or with older data being displayed. The ideal that would apply in this case is "It's better than nothing".
- Data management – As mentioned above, because so many routing scenarios can be digitally compiled so quickly, engineers and designers need to make sure they have a robust data management system. This is not just hardware, but also processes to deal with change management.

8.3 Web 2.0 and the GeoWeb

To say that the internet has improved our ability to find and disseminate information, products, and communicate to the global community would be an understatement. Although what we find on the internet is not always correct or in the best interest of the public, it serves as a global commons for sharing. The internet of recent times, as most of us were originally exposed to, is a vast network of documents, web pages, and images linked by hyperlinks and URLs. Typically referred to as the World Wide Web, it is searchable using a web browser and search engine to retrieve topics on every subject indexed by key words and phrases. We can read, publish, communicate, and shop online.

As with any technology, culture, or movement the internet is constantly undergoing evolution and it is certainly on the forefront of change. Current trends fall under the description of Web 2.0. This is an environment that engages and enables the community far more that the earlier internet has done. Although the internet has broken down major barriers to communicating and publishing information and ideas, it still requires a level of computer skills (e.g. building and maintaining web pages) not shared by the majority of society.

What Web 2.0 offers is an environment more conducive to collaboration, sharing, and e-commerce, without much need for anything more complicated than an internet connection. Web 2.0 is a concept founded on reversing the traditional structure of the internet where content providers pushed their information out to the masses. Instead, things such as YouTube, MySpace, Wikipedia and blogging have provided the conduits through which communities can collaborate on what is pushed out to the web. Software like Linux, SharePoint, and Firefox are examples of how the internet is changing to facilitate collaboration and sharing of information as opposed to one-way communication. Examples like streaming video, podcasts, peer-to-peer exchanges, wiki's, mobile computing and e-commerce further demonstrate globally available tools for the user community to share and provide internet content. Even internet-based telephony like VoIP (Voice-over-Internet Protocol) and Skype have impacted our ability to stay connected with each other using new business models.

What does all this mean for the pipeline industry? We are clearly starting to see the benefits of many of these changes in our day-to-day lives but how does the geo-spatial world connect to the Web 2.0? There are two main contributions emerging, Web Mapping Services (WMS) and the GeoWeb.

8.3.1 Web Mapping Services

Over the last few years there have been major improvements in the area of systems interoperability. Of particular note is the development of Web Services. Essentially, a Web Service is a Web-based software application that is used by another software application. That is, it is a self-contained computer system whose functionality and data can be leveraged by another computer system, even over great geographical and corporate boundaries, and across heterogeneous computer operating systems, development environments and networks.

The implications of Web Services within the context of Web GIS are vast. By being able to access the functionality and data of various systems, the need to implement functionality or acquire data is removed. For example, a government agency may expose vital geographical transportation infrastructure or vegetation data via a Web Service. It is then possible to create a system that can access this Web Service in-house, even if the systems are running on two different operating systems, and developed in two different development environments.

By making data available this way, people and organizations hosting and serving data have begun to bring down a larger potential to effectively use GIS, that is, data and accessibility. By being able to link to data that we previously had to download and maintain is helping us focus on our engineering and not data maintenance. This is the efficient and responsible division of labor that comes from a well designed system, data companies look after data and engineers look after engineering.

At the heart of web map services is the concept is distributed geospatial analysis. By linking geospatial data stores together through the internet or intranet, the requirement for local data hosting and management is removed. Increasingly, GIS tasks and analyses previously performed within specialized GIS software applications locally installed are now being replaced by highly customized web GIS. For years GIS software applications have been refined to become very powerful and comprehensive tools as the boundaries of what is expected continuously gets pushed. This has traditionally resulted in the need for highly specialized GIS analysts to interact with the complex GIS applications and databases. Every time a new type of geospatial workflow or analysis is conducted, a great deal of time spent on figuring out the process. Once the process is figured out, it could be duplicated over and over again with minimal effort, which of course is one of the key benefits of GIS. Nonetheless, as more and more processes are defined, it becomes apparent that in many instances the tasks required for a particular GIS process only utilize a very small percentage of the full functionality available within a desktop GIS software application. Two problems arise from this situation;

- GIS analysts, like most knowledge workers, become bored and uninspired very quickly in an environment where much of their time is spent running routine analyses and repetitive tasks.
- The person requesting information from their GIS analyst has little or no control over the analysis and is further constrained from getting the information they need based on the GIS analyst's schedule and workload.

Web-enabled geospatial tools are being used to help alleviate these problems. Due to the well defined processes in most geospatial workflows, the sequence of steps for a given task can very easily be programmed into a web-based interface. The benefits of placing a geospatial workflow into a web-based format are many. First, it allows anyone with an internet connection and standard web-browser access to geospatial tools. This alleviates the need for GIS analysts to perform repetitive tasks and allows geospatial functionality to be deployed out to many individuals without the need for complex local GIS. Second, it extends the power of GIS to non GIS experts. An engineer can now gain access to specialized and specific geospatial workflows though web-based portals and need not worry about taking the time to learn all of the intricacies involved in operating a complex GIS software application. Moreover, the often CPU intensive processing required for geospatial operations is in most cases conducted remotely on the centralized server that the web-enabled GIS tool is linked to.

In many ways one could argue that we have come full circle and are back in a similar situation to the early days of computing that involved mainframes. Of course the hardware and communications technologies are much more efficient today which supports centralized computing much better than in the past. In any event, this frees up the local machine from having to be highly powerful and allows it to remain dedicated to whatever it's primary function is, which is often not related to geospatial analysis. Web-enabled geospatial applications allow non-geospatial people to focus on what their job is without having to become a GIS expert. They also free up the GIS analysts time so that they can focus on their role of maintaining the back-end system that supports the web-GIS environment and developing new geospatial processes to assist the business in fulfilling its mission. Ultimately, web-based geospatial deployment creates a foundation for that effective division of labor that as was discussed earlier where the various disciplines can focus on being the very best that they can within their own discipline.

In a pipeline business model where so many of the tasks and workflows follow well established processes, either because of strict regulatory requirements or because of the engineering principles and the physics of pipeline design and construction, web-enabled geospatial infrastructure can be immensely valuable. Consider a least cost path exercise similar to that discussed in the virtual Earth section. In many cases the cost elements that need to be considered in choosing a route option will be standardized. Thus the workflow very quickly becomes a specific sequence of geospatial analyses that do not change. The environmental variables might change but the costs elements do not. This kind of exercise can then easily be written into a web-based application and given to the engineer to run as many times as he wishes. Regulatory types of requirements do not change so once their unique requirements are built into the web-based tool, the engineer for example can change the environment variables as many times as he wants and run scenario after scenario without having to be a geospatial expert.

There are many technologies currently in use for the design and development of web services. Most of these are based on traditional approaches to programming and software development. A comprehensive discussion of these is outside the scope of this book however, there are a few relatively new concepts that are gaining widespread acceptance and having a sizeable impact on the proliferation of web services. One is open source programming and the other is cloud computing. Both of these concepts are part of the new Web 2.0 paradigm of collaboration and community involvement that rely heavily on the internet.

Open source programming while not a new concept in general, is becoming an increasingly popular model for geospatial web services development. Founded on the concept of completely open source code for software, it is an approach to software design development and deployment that has the potential to yield large cost savings. Geospatial development, like any other information system, is under constant pressure reduce costs. Open source can be an effective method of mitigating development costs.

Cloud computing on the other hand, is for the most part a new approach to the consumption or use of software. It is centered on the idea of using and paying for software only on an as needed basis. Under this model, software is accessed via the internet and only what gets used is what gets paid for. No software is installed locally and all processing is done remotely on the servers where the software resides. This approach to software usage negates the need for full ownership, maintenance and updates. This is a very new practice and has not seen wide proliferation as yet. Given the direction that geospatial technology has already taken with respect to web services cloud computing holds great promise for Geomatics.

8.3.2 The GeoWeb

Web services are taking us in an exciting direction by enabling access to data and programs that are hosted and maintained by groups who are best suited to do so. Distributing the storage and maintenance of data is truly a smart solution given that peta-bytes of data will overwhelm us as it continually finds its way onto the internet and into markets. Web services and data management may help us deal with volumes of ever changing data, but the GeoWeb is definitely the next exciting step.

One of the most significant challenges facing those who work with geospatial data is finding a way to search for information based on location. For better or worse, the internet provides the architecture for storing and retrieving this data. On the one hand, nothing else has the same capacity to support such a vast network of data, users, and processes. On the other hand, the internet has some serious limitations, particularly in it's search capabilities. While search engines have made many improvements in this area, the fact remains that text-based searches are inherently limited. When we enter a query, a search engine retrieves relevant results based on its algorithm. In some cases spatial information is available, such as when searching for a restaurant and the search results might include a map showing the location of the search result. Google Earth is similar in that it displays photo links, and features of interest based on their location. However, this spatial information is limited to displaying results (sometimes very inaccurately), not searching. For example, it is not possible with traditional internet search capabilities to select a geographic region and ask "show me where all the restaurants are in this area", even when the information is available.

This is where the GeoWeb comes in; it makes all spatially-located information available through a GeoWeb browser, like the one shown in Figure 1. With the browser, you can specify geographical criteria (such as by clicking on a map) and use the geographical criteria to search for information, either alone or in conjunction with traditional text-based searches. For example, a regulator could post information on pipelines and facilities and anyone could search the map and ask it questions like "show me all pipe greater than 12 inch diameter within 10 kilometers of my house". This may sound like a standard GIS query, which it is. However, the dramatic difference is that due to the nature of the internet, data is available for you to query from anyone or anywhere. Therefore, you could do your pipeline search, post an ad selling your car (showing potential buyers where you are located), and search for local Italian restaurants all from the same system. People and organizations are serving up data and information in a vast seamless matrix, a network of servers and peer-to-peer connections that put data at your fingertips and allow you to collect, distribute, communicate, and analyze data and information based on where it is (2).

Without the GeoWeb being fully developed and commonly used, it is difficult to accurately say where the major advances in our pipeline industry will occur. However, we can summarize some notable benefits from which we can infer or predict the changes we will likely see. These benefits include (3):

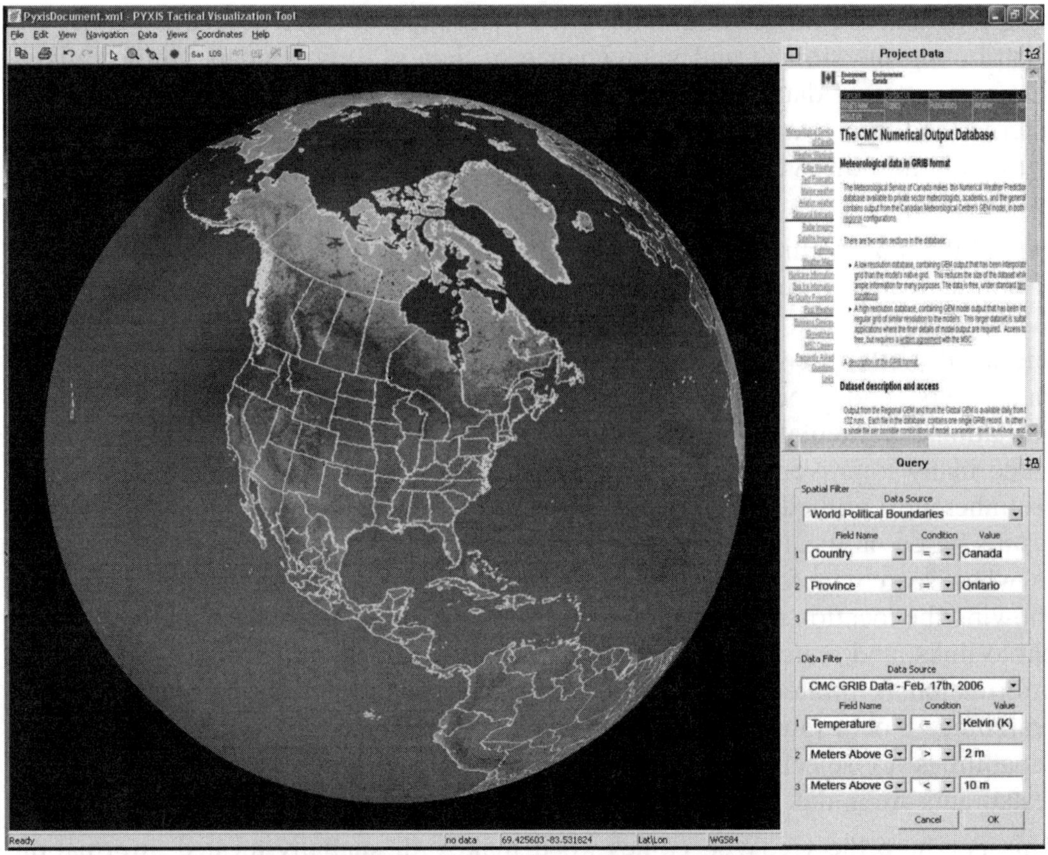

Figure 1. An example of a GeoWeb browser showing the global navigation ability as well as the text search capabilities

- Intuitive indexing and searching – Nearly everyone relates to objects in some way based on where they are. Spatial indexing and searching is an evolution in exploring internet content.
- Open and accessible environment – No one company or institution can hope to index all geographically related information for the whole planet. Therefore, the GeoWeb is an opt-in scheme where any user can register their own data, be it information about their supermarket, or as-built information directly from the field.
- Integrate disparate data sources – Instead of requiring a service provider to compile, correct, convert, and manage volumes of different data sets, the GeoWeb allows users to search, discover, and integrate disparate sources of information (including your own managed data) through the GeoWeb browser software and the global network of published information. All geospatial data will be synchronized on a common framework or grid to facilitate integrated multi-source use.

- Massively scalable architecture – The entire GeoWeb index could be stored on as many servers and peer-to-peer networks as the system wants. It is organic, open, and seamless.
- Index any content – The GeoWeb can index any content, not just HTML. Movies, sound files, text, 3D models, terrain, can all be indexed. Content such as HTML with geographic meta tags allows an easy way to register that content in the GeoWeb. This aspect will provide the core capabilities of a Digital Earth – using location as an organizational structure for all information.
- Built-in security – Security is imperative to ensure that private data remains private, commercial data is purchased, while still providing easy access to public information. In addition, validation and endorsement schema are proposed to allow appropriate organizations to stamp content as being authoritative.
- Support high bandwidth demands – Massive distribution of the geographic database enables many clients around the world to retrieve data simultaneously and at high speed.
- Local control – Individuals, companies, institutions, or local governments can manage their servers and determine what is available to the community.
- Powerful capabilities of a GeoWeb Browser, including:
    - Display – extensible Virtual Globe
    - Search – Geospatial search "what is here" over multiple data sources and services.
    - Data management – like Windows Explorer except geospatial relevant information
    - Data Transformations – Data Fusion, Analysis and Modeling
    - GIS, Image Processing and Database Query capabilities across multiple sources
    - Publishing – as a service, multisource, sensor web, etc.
    - Application Development – Able to build vertical applications on GeoWeb Platform

The GeoWeb concept is not brand new. Al Gore actually dedicated an entire speech to the idea in 1998 (4). What is exciting is the prospect of how it will change things in our industry. With pipelines being such spatially consuming objects, we will find a powerful friend in the GeoWeb.

8.4 Laser Scanning

Once in a while something comes along that really helps us do our work better than before with more control, accuracy, and at less cost. This is what laser scanning has done for many parts of the pipeline lifecycle. Laser scanning actually consists of two main types: Airborne LiDAR (Light Detection And Ranging) and high definition (HD) scanning, which is ground based. Although both these technologies are not particularly new and are already in use on pipeline projects, their widespread adoption has not yet occurred possibly due to limited understanding of the data and technologies, the intimidating size of the data files, difficulties incorporating it into current workflows, or bad past experiences. Hopefully, this summary will provide some unbiased and practical information to help the adoption of laser scanning into your projects and workflow.

8.4.1 Airborne LiDAR

Within the last decade, the airborne LiDAR service offering has evolved to the point where it can provide accurate ground surface elevations on a dense grid (often 1m by 1m) along pipeline corridors, at a cost that is a fraction of the cost for a comparable ground based topographic survey. Airborne LiDAR surveys provide a rapid and fully automatic method of defining the topography of the ground surface over relatively large areas. The method is also significantly faster and more accurate than can be achieved from stereo airphotos using conventional photogrammetry.

To fly a mission, a laser scanner is fitted to either a fixed wing aircraft or a helicopter. The scanner emits a laser pulse that travels to a point on the ground surface that reflects the pulse back to the scanner. The time between the emission and return of the laser pulse is measured so that the distance from the scanner to the point on the ground surface can be calculated since the velocity of light is known. By integrating the laser with a GPS and inertial measurement unit, accurate ground and elevation coordinates (typically ±30cm) are computed. Since there are laser returns from the vegetation cover as well as the ground, filtering algorithms are applied to remove the vegetation cover and provide a bare-earth model (Figure 2).

The accuracy and density of topographic data acquired by LiDAR surveys can be used to greatly facilitate pipeline design and reduce pipeline construction costs. The construction and operating costs, as well as the environmental impacts of pipelines can be minimized by avoiding sloping ground, unstable slopes, poorly drained areas, and other difficult ground conditions. In the past, the selection of optimum pipeline routes in remote areas has relied on the use of stereo airphotos. In recent years, LiDAR surveys have been successfully used to supplement and in some cases replace, conventional airphotos for pipeline route studies (5). Furthermore, many LiDAR systems are integrated with a digital camera and other sensors to collect a variety of data layers.

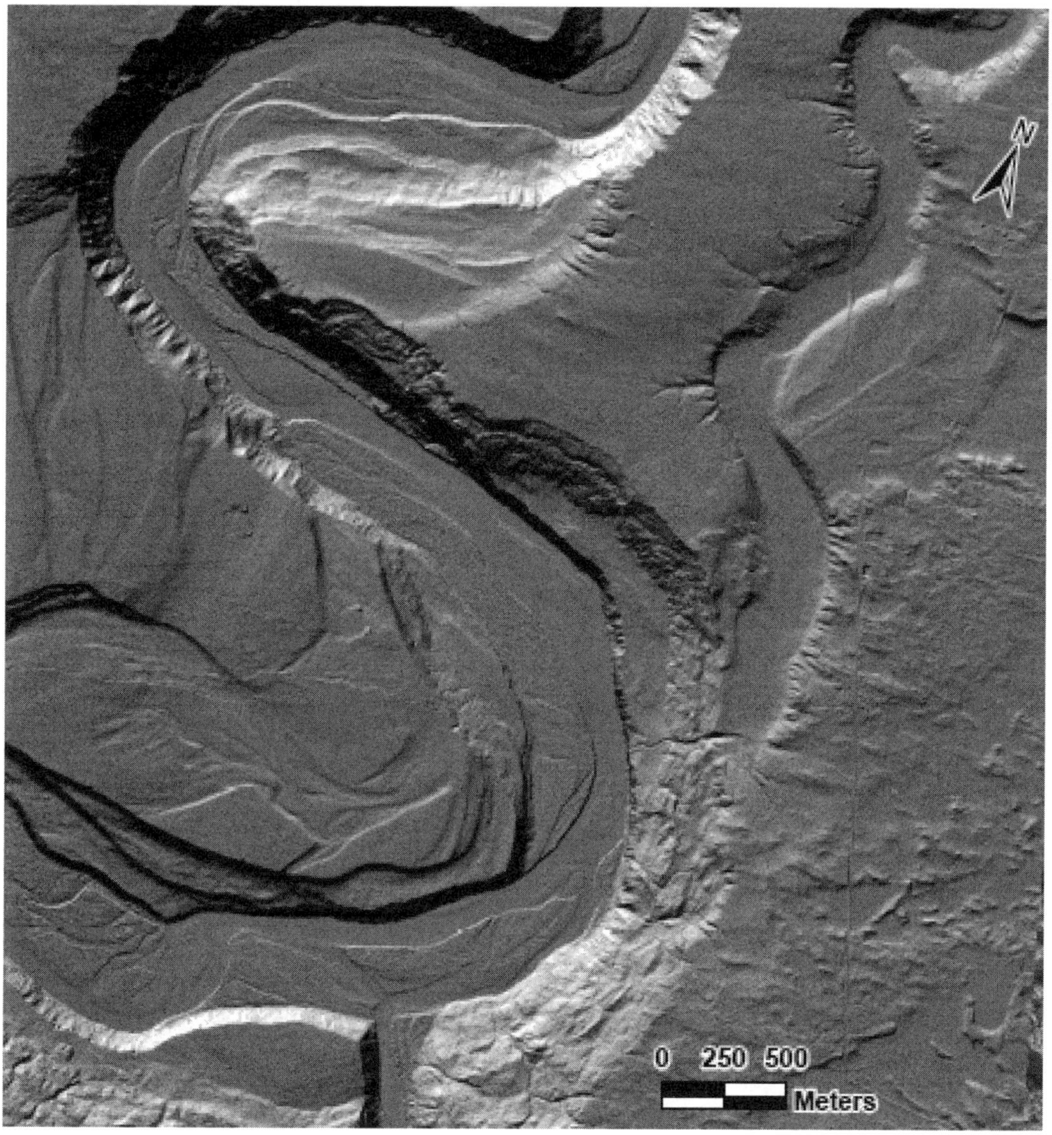
N
0 250 500
Meters

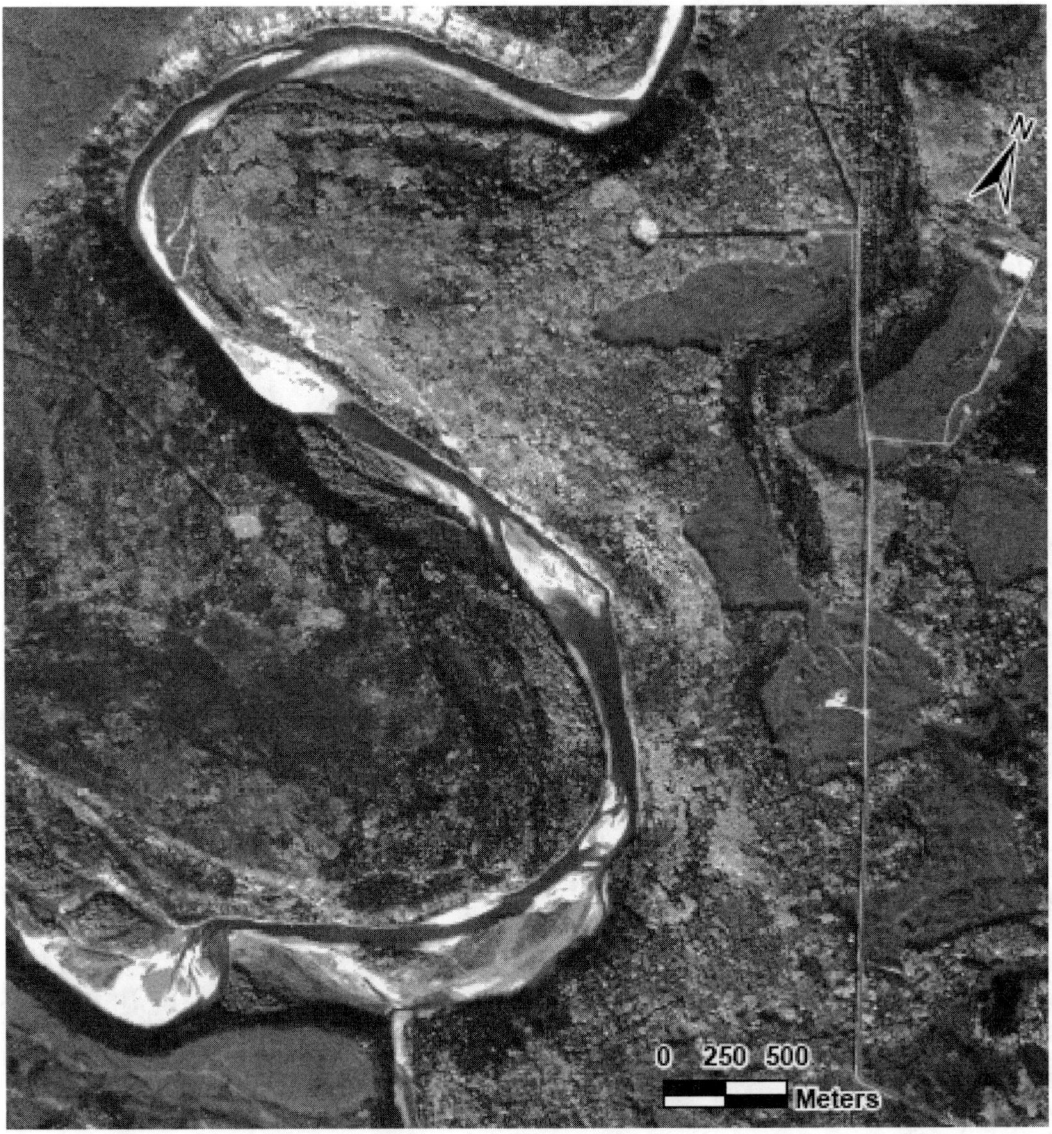

Figure 2. Comparison from of a hillshade relief map (top) with the corresponding air photo (bottom). It is very clear that topography is easier to see in the hillshade than in the airphoto.

The incorporation of LiDAR data into a workflow is a hurdle that typically needs to be overcome. The bigger barriers are mainly due to large file sizes and the limited knowledge of different data products can be derived from LiDAR besides just contours. Therefore, we should address the two problems; large file sizes, and the analyses and manipulations that LiDAR supports to solve engineering problems.

8.4.2 Data density

LiDAR data is often collected and delivered to a grid point density of 1m x 1m. This provides a very realistic depiction of the vegetation covered as well as the bare ground. It can also result in files that are too large for traditional CAD programs to deal with. There are two solutions for this problem:

- Wait for CAD systems to improve and be able to deal with this data, or use GIS to thin out the LiDAR data for CAD use. This is not a great solution since it negates a major benefit of LiDAR, that is, high data density.
- Use GIS instead of CAD to create drawings, maps and reports since the architecture of GIS and the way it reads and manipulates files makes it more efficient at dealing with LiDAR data.

8.4.3 LiDAR analyses

Because LiDAR is dense digital data, topographic and morphological analyses can be performed. These outputs can be used to streamline pipeline routing or monitoring routines. Some examples include:

- Digital surfaces depicting slope and aspect can be created from LiDAR data identifying very explicitly cross and long-slopes to avoid or slopes requiring stabilization.
- Locate water crossings, the class of crossing, width of wetted channel, and crossing quantities
- Pipe upheaval and displacement, trench and backfilling specifications
- Right-of-way preparation, workspace locations, cut/fill quantities, right-of-way construction options

A treatment of routing using digital desktop techniques is outlined in Chapter 3 and also covered in a recent ASME publication Pipeline Geo-Environmental Design and Geohazard Management (6).

8.4.4 High Definition Scanning

Ground-based high definition (HD) scanning is conceptually similar to airborne LiDAR in how a laser is emitted, reflected, and the time delay is measured. HD however, collects data at orders of magnitude greater than LiDAR, thousands of points per square meter versus less than 5 points per square meter for LiDAR. This is due to equipment and the proximity of the sensor to the feature being imaged, where LiDAR is 1000m away, HD can be 100m from an object. You could imagine that the laser actually 'paints' the feature with laser strikes, resulting in a very fine point cloud of the feature (Figure 3). The point cloud is then entered into a 3D software design package to create the desired output.

Figure 3. 3D laser scan of pipe with the corresponding digital photo overlaid onto the point cloud (Image courtesy of Point Geomatics)

By capturing such dense data, a particular feature or event is accurately recorded for desktop analysis and design. This saves time in the field and is safer since it reduces the exposure that a surveyor or engineer is in a dangerous field situation. The realism generated by laser scanners has even gained popular television coverage in shows like Crime 360, where crime scenes are recorded in HD and then critically analyzed for evidence. With ever-growing computer power, digital storage capabilities, and decreasing cost of equipment and services, HD will become a mainstream tool for pipeline and facility data collection.

These systems are typically used for imaging facilities, although emerging mobile methods (mounted on vehicles) are being tested and could be applied to larger applications like right-of-way investigations. The applications that are emerging as standard practices for pipeline and facility design include:

- Establish "as-found" locations for retro-fit work on existing pipeline facilities
- Obtain measurements in areas that are difficult to access or unsafe
- As-building existing facilities
- Reverse engineering structures that have no design plans
- Accident investigation
- Monitoring and engineering side slopes against potential failure
- Civil works, cut and fill, and changes in site morphology.

Although this technologies use is only starting to gain attention for pipeline work many of the benefits experienced in other disciplines include:

- Rapid data collection turnaround
- Highly accurate (millimetres)
- Reduced shut down time due to the speed of data collection
- Safer and less obtrusive data capture
- Reduced engineering and construction costs
- Better retro-fit which leads to dramatic reductions in re-work
- Increased factory fabrication and reduced field fabrication

8.5 Closing Remarks

In the beginning of this book we set out some goals which we want readers to experience. These include some common topics such as the technical aspects of geomatics and their applicability to the pipeline lifecycle. We also cast a glance into the future and outline some topics that illustrate geomatics' continued impact on pipeline engineering and operations.

Geomatics has brought spatial-thinking to an inherently spatial phenomenon. Pipelines, by their very nature, cover geographic space because that is why we build them…to carry products efficiently, from one place to another, over short and long distances. Furthermore, the amount of digital data generated by project engineering and integrity and environmental programs is growing at a rate that causes traditional file-based systems to become overwhelmed. By providing effective methods of collecting geospatial data and efficient ways of storing these data using their location as an index, geomatics evokes innovation in pipeline business.

One last goal that is important to the compilation of this book, but may not be apparent to the reader, is the collaborative, multi-disciplinary approach with which the book is written. Using collaborative technologies such as SharePoint to bring disparate professionals together and collaborate on writing a book over a short period of time supports many of the ideals we present in this book. The collaborative, multi-disciplinary writing approach mirrors the same business approach required in designing, building, and operating pipelines in the modern digital and enviro-socially conscious age. Web and other information technologies are enabling us to easily collaborate and retain the fruits of our collaboration. Spatial data management systems are another enabling technology that helps us retain not only data we collect on a pipeline project but knowledge as well. With the majority of pipeline engineers retiring in the next ten years, this type of knowledge retention is critical to maintaining pipeline engineering and business practices.

Finally, by highlighting geomatics in all major components of the pipeline business cycle we hope we have shown readers the natural fit between these two great disciplines. We recognized early on in planning this book that printed text becomes dated quickly. As a result, our aim is to provide fundamental hubs of information from which readers can explore further sources and successfully implement geomatic technologies and practices to their work. Innovation is about change and the more we know about what is changing around us, the better prepared we are to reap its rewards.

References

(1) Eggers, W. D., Startup, T., "Closing the Infrastructure Gap: The Role of Public-Private Partnerships," Deloitte Development LLC, 2006

(2) Scharl, A., Tochtermann, K., "The Geospatial Web: How Geobrowsers, Social Software and the Web 2.0 are Shaping the Network Society (Advanced Information and Knowledge Processing)," Springer-Verlag, London, 2007

(3) Adapted from http://www.ai.sri.com/dgeo/advantages.shtml

(4) The speech can be seen at http://www.pyxisinnovation.com/Documents/The_Digital_Earth_Understanding_Our_Planet_in_the_21st_Century.pdf

(5) Smith, B, Wong Y., Adam S., "The use of LiDAR to facilitate design and construction of pipelines," Proceedings of the 7th International Pipeline Conference (IPC 2008), Calgary, Canada, ASME, 2008

(6) Rizkalla, M., "Pipeline Geo-Environmental Design and Geohazard Management," ASME, New York, 2008

Glossary

Aerial imagery	Digital images acquired from an airplane or a helicopter used for mapping applications
Aquifer	an underground geological formation of water-bearing permeable rock or unconsolidated materials (gravel, sand, silt, or clay) that contains or conducts groundwater, especially the one that supplies water for wells or springs.
API	American Petroleum Institute (API) is the main U.S trade association for the oil and natural gas industry, representing about 400 corporations involved in production, refinement, distribution, and many other aspects of the industry
Anthropogenic	caused or produced by humans.
AMS	Asset Management System (AMS) – A software application used to manage an organizations physical assets and infrastructure.
Attributes	Characteristics about a spatial feature
Automated Alignment Sheets	Alignment sheets generated by querying a database that stores information occurring along the route. A template instructs the query to extract and place pipeline events in a pre-define window and labels features accordingly.
ACCC	Australian Competition Consumer Commission (ACCC) is an independent authority of the government of Australia. It was established in 1995 with the amalgamation of the Australian Trade Practices Commission (TPC) and the Prices Surveillance Authority to administer the Trade Practices Act 1974 (Cth). Its mandate is to protect consumer rights, business rights and obligations, perform industry regulation and price monitoring and prevent illegal anti-competitive behaviour.
AER	Australian Energy Regulator (AER) The Australian Energy Regulator enforces these rules that the Australian energy market is to abide by.

CEPA	Canadian Energy Pipeline Association (CEPA) represents Canada's transmission pipeline companies. Our members are world leaders in providing safe, reliable long-distance energy transportation. CEPA is dedicated to ensuring a strong and viable transmission pipeline industry in Canada in a manner that emphasizes public safety and pipeline integrity, social and environmental stewardship, and cost competitiveness.
Cartography	The discipline of designing and making maps
Cathodic Protection	A technique by which underground metallic pipe is protected against deterioration (rusting or pitting)
Central Database	A central data repository that is managed with standards and processes to ensure proper change management, security, and data integrity
Change Management	A set of processes put in place to support changes in data or information within a project or enterprise. These processes keep everyone organized and reduce the risk of using incorrect data versions
CRM	Client Resource Management System (CRM) - The system a company uses to track and organize its contact with its current and prospective customers.
Close Interval Survey (CIS)	An inspection technique that includes a series of above-ground pipe-to-soil potential measurements taken at predetermined increments along the pipeline and used to provide information on the effectiveness of the cathodic protection system.
Cloud computing	is a style of computing in which dynamically scalable and often virtualized resources are provided as a service over the Internet. Users need not have knowledge of, expertise in, or control over the technology infrastructure in the "cloud" that supports them.
CAD	Computer-Aided Design (CAD) – A system used by architects, drafters, engineers, to create design and construction drawings.
Connectivity	Describing how arcs are connected by recording the from and to nodes for each arc. Arcs that share a common node are connected.

Consequence	The impact that a pipeline failure could have on the public, employees, property and the environment
CSA	Core Security Area (CSA) is areas that give wildlife a refuge from people for a period of time.
Cumulative Effects	the impact on the environment which results from the incremental impacts of action when added to the other past, present, and reasonably foreseeable future actions regardless of what agency or person undertakes such other actions.
Data-centric	Often applied to modern approaches to engineering where design leverages the wealth of digital data available to analyze and model phenomena
Data model	A design process for a database which identifies and organizes data in tables
Datum	A mathematical surface on which a mapping and coordinate system is based
DEM	Digital elevation models are grids where each cell value represents the elevation on the earth's surface.
DMS	Document Management System (DMS) - A computer system used to track and store electronic documents and/or images of paper documents.
DRASTIC	A methodology developed by the USA EPA in 1987 that would allow the pollution potential of any hydrogeologic setting to be systematically evaluated anywhere in the United States and has been adopted in many other countries.
EDM	Electronic Distance Measurement – Part of the total station surveying instrument used to read distances to a particular point.
EIA	Environmental Impact Assessment (EIA) – An assessment of the possible impacts – positive or negative – that a proposed project may have on the environment; considering natural, social and economic aspects.
ESRI	Environmental Systems Research Institute is an organization that develops widely used GIS software.
EPCM	Engineering, Procurement, Construction Management
ERCB	Energy Resources Conservation Board (ERCB) is an independent, quasi-judicial agency of the Government of Alberta. It regulates the safe, responsible, and

	efficient development of Alberta's energy resources: oil, natural gas, oil sands, coal, and pipelines.
ERP	Enterprise Reporting Planning (ERP) – The system used to integrate many of an organizations back office systems including planning, manufacturing, maintenance, distribution, shipping, and accounting.
FERC	Federal Energy Regulatory Commission (FERC) is the United States federal agency with jurisdiction over interstate electricity sales, wholesale electric rates, hydroelectric licensing, natural gas pricing, and oil pipeline rates.
Failure Likelihood	The relative measure of the likelihood of the pipeline failing as a result of a design or operating condition.
Gas	Any gas or mixture of gases suitable for domestic or industrial fuel and transmitted or distributed to the user through a piping system. The common types are natural gas, manufactured gas, and liquefied petroleum gas distributed as a vapor, with or without the admixture of air.
Geodatabase	A database used to hold a collection of GIS datasets
Geographic Coordinate System	One of the most common coordinate systems which uses degrees of latitude and longitude to describe a location on a spherical version of the earth's surface.
Geomatics	The science of examining the spatial world
Geodetic/Geodesy	Defines the size and shape of the earth and the origin and orientation of the coordinate systems used to map the earth.
Geometric networks	A network of geospatial features that are linked to each other through a common purpose or relation.
GeoPIG	An inline inspection tool that uses a GPS for positioning.
Geoprocessing	Computer programs or models which perform operations on geospatial data
Geospatial	Referring to an objects location or position on the earth or in reference to some datum
Geotiff	An image in TIF file format with geo-referencing information in the header
GeoWeb	Merging of geographical (location-based) information with the abstract information that currently dominates

	the Internet. This would create an environment where one could search for data and information based on location instead of by keyword only
GIS	Geographic Information System, a system of computer software, hardware, data, and personnel to help manipulate, analyze, and present information that is tied to a geographic location
GPS	Global Positioning System, a constellation of satellites developed by the US Department of Defense used for navigation by the military and civilians
Groundwater	water located beneath the ground surface in soil pore spaces and in the fractures of lithologic formations
HSI	Habitat Suitability Index (HSI), a relative index that evaluates the relative suitability of habitat for different species of wildlife based on vegetation cover, food, relative distance to water bodies and other biotic or abiotic factors that influence where that wildlife species would choose to live
HCA	High Consequence Area (HCA), an area where a pipeline will travel through that is an unusually sensitive area, like a populated area, a protected site (environmental or cultural resource), or commercially navigable waterway
High Definition (HD) scanning	a ground-based laser scanner (similar to aerial LiDAR) that emits a laser pulse at an object and is able to digitally/virtually reconstruct the object based on the return signal from the laser. Used to render pipeline components and facilities for design, as-building, retro-fitting, reverse engineering, and accessing potentially unsafe or awkward work-places
Hillshade	An image derived from an illuminated DEM providing artificial shadows to give a 3D effect
HRM	Human Resource Management (HRM) is the solution used to automate many of the aspects of human resources, including benefits administration, payroll, recruiting and training, and performance analysis and review.
Implementation	Refers to guiding a client from purchase to use of the software or hardware that was purchased. This can include Requirements Analysis, Scope Analysis,

	Customizations, Systems Integrations, User Policies, User Training and Delivery
Innovation	Is a new way of doing something. This may be incremental, or disruptive which is a more radical and revolutionary change in thinking, products, processes, or organizations. A distinction is typically made between invention, an idea made manifest, and innovation, ideas applied successfully.
In-line Inspection (ILI)	A pipeline inspection technique that uses devices known in the industry as "smart pigs". These devices run inside the pipe and provide indications of metal loss, deformation and other defects
LACT	Lease Automatic Custody Transfer unit measures volumes in the range of 100-1000 BOPD. This system provides for the automatic measurement, sampling, and transfer of oil from the lease location into a pipeline. A system of this type is applicable where larger volumes of oil are being produced and must have a pipeline available in which to connect
Land Management System	An information system that keeps track of land owner information, cadastral boundaries, permits, zoning, and land owner agreements
Latitude	A spherical reference system used to measure locations on the earth's surface. Latitude measures angles in a north-south direction
Least Cost Path	Using start and end points for a pipeline route, an algorithm that selects a route with the least amount of friction. Friction is determined by weighting different surfaces based on their suitability to being crossed by the pipeline
LiDAR	Light detection and ranging is a technique that transmits laser pulses to the earth's surface and collects the return signals to map the terrain elevation.
Linear Referencing	A method for associating attributes or events to locations or segments of a linear feature
Linux	is a generic term referring to Unix-like computer operating systems based on the Linux kernel. Their development is one of the most prominent examples of free and open source software collaboration; typically all the underlying source code can be used, freely

modified, and redistributed by anyone under the terms of the GNU, GPL and other free licenses.

Longitude	A spherical reference system used to measure locations on the earth's surface. Longitude measures angles in a east-west direction.
Metadata	Legacy information about a particular dataset, typically including at least the data source, vintage, quality, accuracy, and purpose
MFL Inspection	The Magnetic Flux Leakage tool is the most commonly used in-line inspection tool where a magnetic field is used to detect internal or external corrosion.
Mobile Mapping	GIS functionality on a mobile device such as a tablet PC, hand-held GPS, or even a PDA or cell phone
NEB	National Energy Board (NEB) is an independent federal agency established in 1959 by the Parliament of Canada to regulate international and interprovincial aspects of the oil, gas and electric utility industries.
NPMS	National Pipeline Mapping System (NPMS) is a geographic information system (GIS) created by the U.S. Department of Transportation, Pipeline and Hazardous Materials Safety Administration (PHMSA), Office of Pipeline Safety (OPS) in cooperation with other federal and state governmental agencies and the pipeline industry. The NPMS consists of geospatial data, attribute data, public contact information, and metadata pertaining to the interstate and intrastate gas and hazardous liquid transmission pipelines, liquefied natural gas (LNG) plants, and hazardous liquid breakout tanks jurisdictional to PHMSA.
NGO	Non-Governmental Association (NGO) is a non-profit and non-governmental organization created by individuals or groups of individuals with no participation or representation of any government who bring public concerns to governments, advocate and monitor policies and who provide analysis and expertise with respect to public issues including the environment, health or social issues.
OPS	Office of Pipeline Safety (OPS) The Office of Pipeline Safety is the Federal safety authority for ensuring the safe, reliable, and environmentally sound operation of the Nation's pipeline transportation system.

Open source programming	software developed in a public, collaborative manner that permits users to use, change, and improve the software, and to redistribute it in modified or unmodified forms
Ortho-photography	Digital images created from air photos from which camera and topographic distortions have been removed
Paradigm	Historian of science Thomas Kuhn gave this word its contemporary meaning when he adopted it to refer to the set of widely accepted practices that define a scientific discipline during a particular period of time
PFD	A process flow diagram is a diagram commonly used in chemical and process engineering to indicate the general flow of plant processes and equipment. The PFD displays the relationship between major equipment of a plant facility and does not show minor details such as piping details and designations
P&ID	Piping and instrumentation diagram is a schematic diagram showing piping, equipment and instrumentation connections within process units in oil refineries, petrochemical and chemical plants, natural gas processing plants, power plants, and other similar plants
Photogrammetry	The technique of determining the position and shape of objects from stereo aerial photography or imagery
PIG	See In-line Inspection
Pipeline	All parts of physical facilities through which product moves in transportation, including pipe, valves, fittings, flanges (including bolting and gaskets), regulators, pressure vessels, pulsation dampeners, relief valves, and other appurtenances attached to pipe, compressor units, metering stations, regulator stations, and fabricated assemblies. Included within this definition are transmission and gathering lines, transporting product from production facilities to onshore locations and storage equipment of the closed pipe type, which is fabricated or forged from pipe or fabricated from pipe and fittings.
PHMSA	Pipeline and Hazardous Material Safety Administration (PHMSA) was created under the Norman Y. Mineta Research and Special Programs Improvement Act (P.L. 108-426) of 2004. The purpose of the Act was to

	provide a more focused research organization and establish a separate operating administration for pipeline safety and hazardous materials transportation safety operations
Pipeline Inspection	The use of a non-destructive testing technique
Pipeline Segment	A length of pipeline or part of the system that has unique characteristics in a specific geographic location
Pixel	A cell of a raster
Polygon	A vector file that consist of at least 3 points which contain an area
PPP	Public-Private Partnership (PPP) describes a government service or private business venture which is funded and operated through a partnership of government and one or more private sector companies. These schemes are sometimes referred to as PPP, P3 or P^3.
Probability	The likelihood of an incident occurring (see Failure Likelihood)
Projections	A mathematical calculation transforming the three dimensional surface of the earth to a two dimensional plane
Projected Coordinate System	See Projections
Proximity	How close two objects are to each other, a spatial phenomena that is powerfully analyzed using GIS.
Pump Station	A facility for pumping fluids from one location to another
QA/QC	Quality Assurance/Quality Control – processes to ensure products or services are designed and produced to meet or exceed customer requirements.
Query	Selecting features in a GIS or database by asking a question or a logical expression
Raster	Contains rows and columns of pixels that contain information such as elevation value.
Resolution	The amount of detail found in one pixel of the image. For example, an image with one meter resolution means that each pixel in the image represents one square meter on the ground.
RFID	Radio Frequency Identification, an identification tag that stores and/or transmits information when induced by an antenna

Riparian	pertaining to, situated or dwelling on the bank of a river or other body of water
Risk	A measure of potential loss in terms of both the incident probability (likelihood) of occurrence and the magnitude of the consequences
Risk Assessment	A systematic process in which potential hazards from facility operation are identified, and the likelihood and consequences of potential adverse events are estimated. Risk assessments can have varying scopes, and be performed at varying level of detail depending on the operator's objectives.
Risk-based lifecycle	An approach to representing pipeline lifecycle risk where Risk = Probability x Consequence. The probability is measured by the adequacy, implementation, and effectiveness of a company's programs. Consequence is the result if an accident occurs.
Routing	a manual or automated process for defining a pipeline route using maps, topography, and other typically spatial datasets
RTK GPS	Is a real time kinematic global positioning system where a rover collects measurements and single reference station provides the real-time corrections to generate a centimeter level of accuracy.
SCADA System	The Supervisory Control and Data Acquisition (SCADA) System that collects data from various sensors along a pipeline system at pieces of infrastructure, such as valves, and then sends it a central computer which then manages and controls the system.
Services Oriented Architecture (SOA)	Is an enterprise integration methodology that focuses on the creation of a collection of reusable Web services that applications used to communicate and exchange information.
Shape file	Editable spatial file format generated in ESRI's software
Signature species	a specie, typically plant or animal, which is noteworthy or identified within a region or environment
Skype	is a software application that allows users to make voice calls over the Internet (http://www.skype.com/)

SDM	Spatial Data Management (SDM) organizes spatial data to ensure data security, integrity, and effective change management
Stationing	See Linear Referencing
Surface water	water collected on the ground or in a stream, river, lake, wetland, or ocean
Tank Farm	a facility for storage of liquid (petro-) chemicals
TIFF world file	A text file that provides geo-referencing information for an image in TIF format
Transmission System	One or more segments of pipeline, usually interconnected to form a network, which transports products from a gathering system, the outlet of a processing plant, or a storage field to a high- or low-pressure distribution system, a large-volume customer, or another storage field.
Ultrasonic Inspection	A pipe inspection device for ultrasonically inspecting the wall of a pipe to provide a record of pipe wall thickness variation and/or defects or other abnormalities.
UTM	The Universal Transverse Mercator (UTM) coordinate system is a grid-based method of specifying locations on the surface of the Earth.
Value Chain	A chain of activities. Products pass through all activities of the chain in order and at each activity the product gains some value. The chain of activities gives the products more added value than the sum of added values of all activities.
Vector	A geographic object that is represented as a point, line, or polygon.
Vendor	Body outside of the influence of the Company
Virtual Earth	A modern digital atlas accessible through the internet that allows users to navigate to places of interest and post their own data
Visualization	Techniques used to simulate 3D or other natural phenomena by combining different data layers in a digital environment.
Web 2.0	Considered the second generation of internet development and use characterized by greater collaboration, sharing, wikis, blogs, and other hosted services

Web GIS	GIS applications deployed to operate and access over the internet or an intranet.
Wiki	A website that uses wiki software that allows easy creation and editing of any number of interlinked Web pages. Used to create collaborative websites like the well known Wikipedia, intranets and knowledge management systems
Yellow Jacket®	provides external protection for pipe used in the oil and gas and waterworks industries where moderate operating temperatures and good handling capabilities are required. It consists of a continuous sheath of high density polyethylene extruded over a rubberized asphalt adhesive.

Index